Mass Spectrometry Imaging in Food Analysis

Food Analysis & Properties

Series Editor

Leo M. L. Nollet

University College Ghent, Belgium

This CRC series **Food Analysis and Properties** is designed to provide a state-of-art coverage on topics to the understanding of physical, chemical and functional properties of foods: including (1) recent analysis techniques of a choice of food components; (2) developments and evolutions in analysis techniques related to food; (3) recent trends in analysis techniques of specific food components and/or a group of related food components.

Flow Injection Analysis of Food Additives
Edited by Claudia Ruiz-Capillas and Leo M. L. Nollet

Marine Microorganisms
Extraction and Analysis of Bioactive Compounds
Edited by Leo M. L. Nollet

Multiresidue Methods for the Analysis of Pesticide Residues in Food
Edited by Horacio Heinzen, Leo M.L. Nollet, and Amadeo R. Fernandez-Alba

Spectroscopic Methods in Food Analysis
Edited by Adriana S. Franca and Leo M.L. Nollet

Phenolic Compounds in Food
Characterization and Analysis
Edited by Leo M.L. Nollet, Janet Alejandra Gutierrez-Uribe

Testing and Analysis of GMO-containing Foods and Feed
Edited by Salah E. O. Mahgoub, Leo M. L. Nollet

Fingerprinting Techniques in Food Authenticity and Traceability
Edited by K.S. Siddiqi and Leo M.L. Nollet

Hyperspectral Imaging Analysis and Applications for Food Quality
Edited by Nrusingha Charan Basantia, Leo M.L. Nollet, Mohammed Kamruzzaman

Ambient Mass Spectroscopy Techniques in Food and the Environment
Edited by Leo M.L. Nollet and Basil K. Munjanja

Food Aroma Evolution
During Food Processing, Cooking and Aging
Edited by Matteo Bordiga, Leo M. L. Nollet

Proteomics for Food Authentication
Edited by Leo M. L. Nollet, Otles, Semih

Mass Spectrometry Imaging in Food Analysis
Edited by Leo M. L. Nollet

Analysis of Nanoplastics and Microplastics in Food
Edited by Leo M. L. Nollet and Khwaja Salahuddin Siddiqi

For more information, please visit the Series Page: https://www.crcpress.com/Food-Analysis--Properties/book-series/CRCFOODANPRO

Mass Spectrometry Imaging in Food Analysis

Edited by
Leo M.L. Nollet

CRC Press
Taylor & Francis Group
Boca Raton London New York

CRC Press is an imprint of the
Taylor & Francis Group, an **informa** business

CRC Press
Taylor & Francis Group
6000 Broken Sound Parkway NW, Suite 300
Boca Raton, FL 33487-2742

Printed on acid-free paper

International Standard Book Number-13: 978-1-138-37069-2 (Hardback)

Visit the Taylor & Francis Web site at
http://www.taylorandfrancis.com

and the CRC Press Web site at
http://www.crcpress.com

Contents

Series Preface

There will always be a need for methods of analyzing food compounds and properties. Current trends in analyzing methods include automation, increasing the speed of analyses, and miniaturization. The unit of detection has evolved over the years from micrograms to pictograms.

A classical pathway of analysis is sampling, sample preparation, cleanup, derivatization, separation, and detection. At every step, researchers are working and developing new methodologies. A large number of papers are published every year on all facets of analysis. So, there is a need for books that gather information on one kind of analysis technique or on analysis methods of a specific group of food components.

The scope of the CRC series on *Food Analysis & Properties* aims to present a range of books edited by distinguished scientists and researchers who have significant experience in scientific pursuits and critical analysis. This series is designed to provide state-of the-art coverage on topics such as

1. Recent analysis techniques on a range of food components
2. Developments and evolution in analysis techniques related to food
3. Recent trends in analysis techniques of specific food components and/or a group of related food components
4. The understanding of physical, chemical, and functional properties of foods

The book *Mass Spectrometry in Food Analysis* is Volume Number 10 of this series. I am happy to be a series editor of such books for the following reasons:

- I am able to pass on my experience in editing high-quality books related to food.
- I get to know colleagues from all over the world more personally.
- I continue to learn about interesting developments in food analysis.

A lot of work is involved in the preparation of a book. I have been assisted and supported by a number of people, all of whom I would like to thank. I would especially like to thank the team at CRC Press/Taylor & Francis, with a special word of thanks to Steve Zollo, Senior Editor.

Many, many thanks to all the editors and authors of this volume and future volumes. I very much appreciate all their effort, time, and willingness to do a great job.

I dedicate this series to

- My wife, for her patience with me (and all the time I spend on my computer).
- All patients suffering from prostate cancer; knowing what this means, I am hoping they will have some relief.

Dr. Leo M. L. Nollet (Retired)
University College Ghent
Ghent, Belgium

About the Editor

Leo M. L. Nollet earned an MS (1973) and PhD (1978) in biology from the Katholieke Universiteit Leuven, Belgium. He is an editor and associate editor of numerous books. He edited for M. Dekker, New York—now CRC Press of Taylor & Francis Publishing Group—the first, second, and third editions of *Food Analysis by HPLC and Handbook of Food Analysis*. The last edition is a two-volume book. Dr. Nollet also edited the *Handbook of Water Analysis* (first, second, and third editions) and *Chromatographic Analysis of the Environment*, third and fourth editions (CRC Press). With F. Toldrá, he coedited two books published in 2006, 2007, and 2017: *Advanced Technologies for Meat Processing* (CRC Press) and *Advances in Food Diagnostics* (Blackwell Publishing—now Wiley). With M. Poschl, he coedited the book *Radionuclide Concentrations in Foods and the Environment*, also published in 2006 (CRC Press). Dr. Nollet has also coedited with Y. H. Hui and other colleagues on several books: *Handbook of Food Product Manufacturing* (Wiley, 2007), *Handbook of Food Science, Technology, and Engineering* (CRC Press, 2005), *Food Biochemistry and Food Processing* (first and second editions; Blackwell Publishing—now Wiley—2006 and 2012), and the *Handbook of Fruits and Vegetable Flavors* (Wiley, 2010). In addition, he edited the *Handbook of Meat, Poultry, and Seafood Quality*, first and second editions (Blackwell Publishing—now Wiley—2007 and 2012). From 2008 to 2011, he published five volumes on animal product-related books with F. Toldrá: *Handbook of Muscle Foods Analysis*, *Handbook of Processed Meats and Poultry Analysis*, *Handbook of Seafood and Seafood Products Analysis*, *Handbook of Dairy Foods Analysis*, and *Handbook of Analysis of Edible Animal By-Products*. Also, in 2011, with F. Toldrá, he coedited two volumes for CRC Press: *Safety Analysis of Foods of Animal Origin* and *Sensory Analysis of Foods of Animal Origin*. In 2012, they published the *Handbook of Analysis of Active Compounds in Functional Foods*. In a collaboration with Hamir Rathore, *Handbook of Pesticides: Methods of Pesticides Residues Analysis* was marketed in 2009; *Pesticides: Evaluation of Environmental Pollution* in 2012; *Biopesticides Handbook* in 2015; and *Green Pesticides Handbook: Essential Oils for Pest Control* in 2017. Other finished book projects include *Food Allergens: Analysis, Instrumentation, and Methods* (with A. van Hengel; CRC Press, 2011) and *Analysis of Endocrine Compounds in Food* (Wiley-Blackwell, 2011). Dr. Nollet's recent projects include *Proteomics in Foods* with F. Toldrá (Springer, 2013) and *Transformation Products of Emerging Contaminants in the Environment: Analysis, Processes, Occurrence, Effects, and Risks* with D. Lambropoulou (Wiley, 2014). In the series *Food Analysis & Properties*, he edited (with C. Ruiz-Capillas) *Flow Injection Analysis of Food*

Additives (CRC Press, 2015) and *Marine Microorganisms: Extraction and Analysis of Bioactive Compounds* (CRC Press, 2016). With A.S. Franca, he coedited *Spectroscopic Methods in Food Analysis* (CRC Press, 2017), and with Horacio Heinzen and Amadeo R. Fernandez-Alba he coedited *Multiresidue Methods for the Analysis of Pesticide Residues in Food* (CRC Press, 2017). Further volumes in the series *Food Analysis & Properties* are *Phenolic Compounds in Food: Characterization and Analysis* (with Janet Alejandra Gutierrez-Uribe, 2018), *Testing and Analysis of GMO-Containing Foods and Feed* (with Salah E. O. Mahgoub, 2018), *Fingerprinting Techniques in Food Authentication and Traceability* (with K. S. Siddiqi, 2018), *Hyperspectral Imaging Analysis and Applications for Food Quality* (with N.C. Basantia and Mohammed Kamruzzaman, 2018), *Ambient Mass Spectroscopy Techniques in Food and the Environment* (with Basil K. Munjanja, 2019), and *Food Aroma Evolution: During Food Processing, Cooking, and Aging* (with M. Bordiga, 2019).

List of Contributors

Nrusingha Charan Basantia
Godfrey Phillips India Ltd
India

Keith E. Belk
Department of Animal Sciences
Colorado State University
Fort Collins, Colorado

Corey D. Broeckling
Proteomics and Metabolomics Facility
Colorado State University
Fort Collins, Colorado

Richard M. Caprioli
Professor of Biochemistry
Director of the Mass Spectrometry Research Center
Vanderbilt University School of Medicine
Nashville, Tennessee

Chandrakala
Faculty of Pharmacy
Maulana Azad University
Jodhpur, Rajasthan, India

Maria Teresa Galceran
Department of Chemical Engineering and Analytical Chemistry
University of Barcelona
Barcelona, Spain

Sadaf Jamal Gilani
Nibha Institute of Pharmaceutical Sciences
Rajgir, Nalanda
Bihar, India

Devin A. Gredell
Center for Meat Safety and Quality
Department of Animal Sciences
Colorado State University
Fort Collins, Colorado

Adam L. Heuberger
Department of Horticulture and Landscape Architecture
Colorado State University
Fort Collins, Colorado

Renata Jandova
Waters Corporation
Wilmslow, United Kingdom

Soo-Young Kim
Department of Statistics
Colorado State University
Fort Collins, Colorado

D. Andy King
USDA-ARS U.S. Meat Animal Research Center
Clay Center, Nebraska

Fangbiao Li
Exploratory Drug Metabolism
Schering-Plough Research Institute
Kenilworth, New Jersey

Domenica Mangraviti
Dipatimento di Scienze Chimiche Biologiche
Farmaceutiche ed Ambientali
University of Messina
Messina, Italy

Nino Marino
Pistì – Antichi Sapori dell'Etna S.r.L Viale J.F. Kennedy/Zona artigianale
Catania, Italy

Luigi Mondello
Dipatimento di Scienze Chimiche Biologiche
Farmaceutiche ed Ambientali
University of Messina
Messina, Italy
University Campus Bio-Medico of Rome
Rome, Italy

Encarnacion Moyano
Department of Chemical Engineering and Analytical Chemistry
University of Barcelona
Barcelona, Spain

Semih Ötleş
Ege University, Faculty of Engineering
Department of Food Engineering
Izmir, Turkey

Vasfiye Hazal Özyurt
Near East University, Faculty of Engineering
Department of Food Engineering
Nicosia, Turkey

Davy Petit
Waters Corporation
Wilmslow, United Kingdom

Jessica E. Prenni
Department of Horticulture and Landscape Architecture
Colorado State University
Fort Collins, Colorado

Michelle L. Reyzer
Education Director
Research Assistant Professor
Department of Biochemistry
Vanderbilt University School of Medicine
Nashville, Tennessee

Francesca Rigano
Dipatimento di Scienze Chimiche Biologiche
Farmaceutiche ed Ambientali
University of Messina
Messina, Italy

Amelia R. Schroeder
Department of Mathematics and Statistics
East Tennessee State University
Johnson City, Tennessee

Raquel Sero
Department of Chemical Engineering and Analytical Chemistry
University of Barcelona
Barcelona, Spain

Steven D. Shackelford
USDA-ARS U.S. Meat Animal Research Center
Clay Center, Nebraska

Julia L. Sharp
Department of Statistics
Colorado State University
Fort Collins, Colorado

Sara Stead
Waters Corporation
Wilmslow, United Kingdom

Tommy L. Wheeler
USDA-ARS U.S. Meat Animal Research Center
Clay Center, Nebraska

Dale R. Woerner
Department of Animal Sciences
Colorado State University
Fort Collins, Colorado

Md. Taleu Zzaman
Faculty of Pharmacy
Maulana Azad University
Jodhpur, Rajasthan, India

Introduction

Mass spectrometry imaging (MSI) is a mass spectrometry technique to visualize the spatial distribution of molecules, such as biomarkers, metabolites, peptides, or proteins, by their molecular masses. The sample is moved to reach all regions until the entire sample is scanned and all mass spectra are collected. By choosing a peak in the resulting spectra that corresponds to the compound of interest, MS data are used to map its distribution across the sample. This results in pictures of the spatially resolved distribution of a compound pixel by pixel. Each data set contains a veritable gallery of pictures because any peak in each spectrum can be spatially mapped. Despite the fact that MSI has been generally considered a qualitative method, the signal generated by this technique is proportional to the relative abundance of the analyte and so quantification is possible.

The most common ionization technologies in the field of MSI are desorption electrospray ionization (DESI) imaging, matrix-assisted laser desorption ionization (MALDI) imaging, and secondary-ion mass spectrometry imaging (SIMS) imaging.

Secondary-ion mass spectrometry is used to analyze solid surfaces and thin films by sputtering the surface with a focused primary ion beam and collecting and analyzing ejected secondary ions.

Matrix-assisted laser desorption ionization can be used as a mass spectrometry imaging technique for relatively large molecules. It has recently been shown that the most effective type of matrix to use is an ionic matrix for MALDI imaging of tissue. In this version of the technique the sample, typically a thin tissue section, is moved in two dimensions while the mass spectrum is recorded.

Desorption electrospray ionization is a less destructive technique, which couples simplicity and rapid analysis of the sample. The sample is sprayed with an electrically charged solvent mist at an angle that causes the ionization and desorption of various molecular species. Then, two-dimensional maps of the abundance of the selected ions in the surface of the sample in relation with the spatial distribution are generated. This technique is applicable to solid, liquid, frozen, and gaseous samples. Moreover, DESI allows the analysis of a wide range of organic and biological compounds, as animal and plant tissues and cell culture samples, without complex sample preparation.

Other MSI techniques used are laser ablation electrospray ionization (LAESI), laser ablation inductively coupled plasma (LA-ICP), and liquid extraction surface analysis mass spectrometry imaging (LESA-MSI).

Applications of MS imaging include analysis of peptides and proteins, lipids, food contaminants, and constituents, and authenticity assessment.

This book contains seven chapters. It starts with a chapter on the fundamentals of mass spectrometry imaging related to food analysis.

Chapters 2 to 6 discuss different MSI techniques such as matrix-assisted laser desorption ionization (MALDI), secondary-ion mass spectrometry (SIMS), desorption electrospray ionization (DESI), liquid extraction surface analysis mass spectrometry imaging (LESA-MSI), and laser ablation inductively coupled plasma (LA-ICP).

The last chapter details different relatively new MSI techniques: laser ablation electrospray ionization (LAESI), laser desorption ionization (LDI), desorption atmospheric pressure photoionization (DAPPI), desorption atmospheric pressure chemical ionization (DAPCI), low-temperature plasma (LTP), and laser ablation atmospheric pressure photoionization (LAAPPI).

This book is the result of hard work and many efforts of the different authors. We give them a warm thank you.

Leo M.L. Nollet

The greatest enemy of knowledge is not ignorance, it is the illusion of knowledge.

Stephen Hawking

SECTION 1

Mass Spectrometry Imaging

CHAPTER 1

Mass Spectrometry Imaging

Leo M.L. Nollet

CONTENTS

1.1 MASS SPECTROMETRY IMAGING

Mass spectrometry (MS) can provide qualitative as well as quantitative information for a wide variety of compound classes. The molecular mass of a compound, calculated from the *m*/*z* value of the corresponding ion, can be used for (tentative) identification. Ion fragmentation experiments (often called 'tandem MS' or 'MS/MS' experiments) provide information about the molecular structure of an analyte. Peak intensities in mass spectra can be calibrated for analyte quantitation (1).

Mass spectrometry imaging combines these capabilities with information on the location of analytes within a sample (1). Mass spectrometry imaging (MSI or IMS) is a very complex analytical technique that analyzes the spatial distribution of molecular species in a variety of samples. It has the capability to image a great number of molecules, such as metabolites, lipids, peptides, or proteins, in a single experiment without labeling (2). The combination of information gained from MS and visualization of spatial distributions in thin sample sections makes this a valuable chemical analysis tool for biological specimen characterization.

After minimal sample preparation, the general setup of an MSI experiment involves defining an (*x*,*y*) grid over the surface of the sample. The mass spectrometer then ionizes the molecules on the surface of the sample and collects a mass spectrum at each pixel on the section with the resulting spatial resolution defined by the pixel size. After collecting the spectra, computational software is used to select an individual mass-to-charge (*m*/*z*) value, and the intensity of the *m*/*z* is extracted from each pixel's spectrum. These intensities are then combined into a heat map image depicting the relative distribution of that *m*/*z* value throughout the sample's surface.

In order to determine the identity of a specific *m*/*z* value, tandem MS (MS/MS) fragmentation can be performed on ions from each pixel, and the fragments can be used to piece together the structure of the unknown molecule. Otherwise, the molecule can be identified based on its intact mass by accurate mass matching to databases of known molecules within a certain mass error range.

The most common ionization technologies in the field of MSI are desorption electrospray ionization imaging (DESI imaging), matrix-assisted laser desorption ionization imaging (MALDI imaging), and secondary ion mass spectrometry imaging (SIMS imaging).

1.2 SAMPLE PREPARATION

Sample preparation is probably the most critical step in MALDI mass spectrometry imaging (3). All procedures should be optimized as to retain the original spatial distribution and molecular identity of analyte compounds in the sample as far as possible. Mass spectrometry imaging experiments are typically performed on (thin) tissue sections. However, classical embedding material for sectioning such as 'optimal cutting temperature' compound (OCT) should be avoided if possible. These polymers often result in substantial ion suppression and background signals which can severely interfere with mass spectrometric analysis. Therefore, cryosections of tissue without embedding are used if possible. More delicate samples can be prepared with embedding materials that have been reported to be compatible with mass spectrometry imaging; these include carboxymethyl cellulose, gelatine, a polymer compound, and tragacanth gum.

Another crucial step in sample preparation is the application of a matrix which is necessary to desorb and ionize sample components. The matrix needs to interact with the sample (surface) in order to extract analytes from the tissue and incorporate them into matrix crystals. At the same time, the spatial integrity of the sample has to be preserved and analyte migration should be limited as far as possible. A number of sample preparation methods have been developed for MALDI imaging. A simple way of matrix application is spraying of a concentrated matrix solution by means of a thin-layer chromatography (TLC) sprayer. This method provides effective analyte incorporation, but can also lead to substantial spatial spreading of analytes if spraying conditions are not carefully controlled. An increasing number of commercial matrix application devices have been developed recently. They include the 'ImagePrep' (Bruker Daltonik, Bremen, Germany) and the 'SunCollect' sprayer (SunChrom, Friedrichsdorf, Germany). Both systems can achieve matrix crystal sizes below 50 μm in diameter.

Spotting devices for MALDI matrix include the 'CHIP-1000' (Shimadzu Corp., Kyoto, Japan) and a spotter from Labcyte (Sunnyvale, CA, USA). These systems provide matrix deposition spots of about 200 μm in diameter.

In Figure 1.1 is represented a schematic of the major stages in sample preparation for MSI. We can distinguish four components: collection, processing, post-sectioning, and ionization (3).

1.3 MICROPROBE AND MICROSCOPE MODE

In microprobe mode IMS a series of mass spectra are recorded at a surface (4). A full mass spectrum is stored for a series of (x,y) coordinates, which could otherwise be visualized as an intensity distribution image at each m/z unit. By coupling all individual mass spectra, a distribution map of intensities as a function of x and y locations can be plotted. As a result reconstructed molecular images of the sample are obtained.

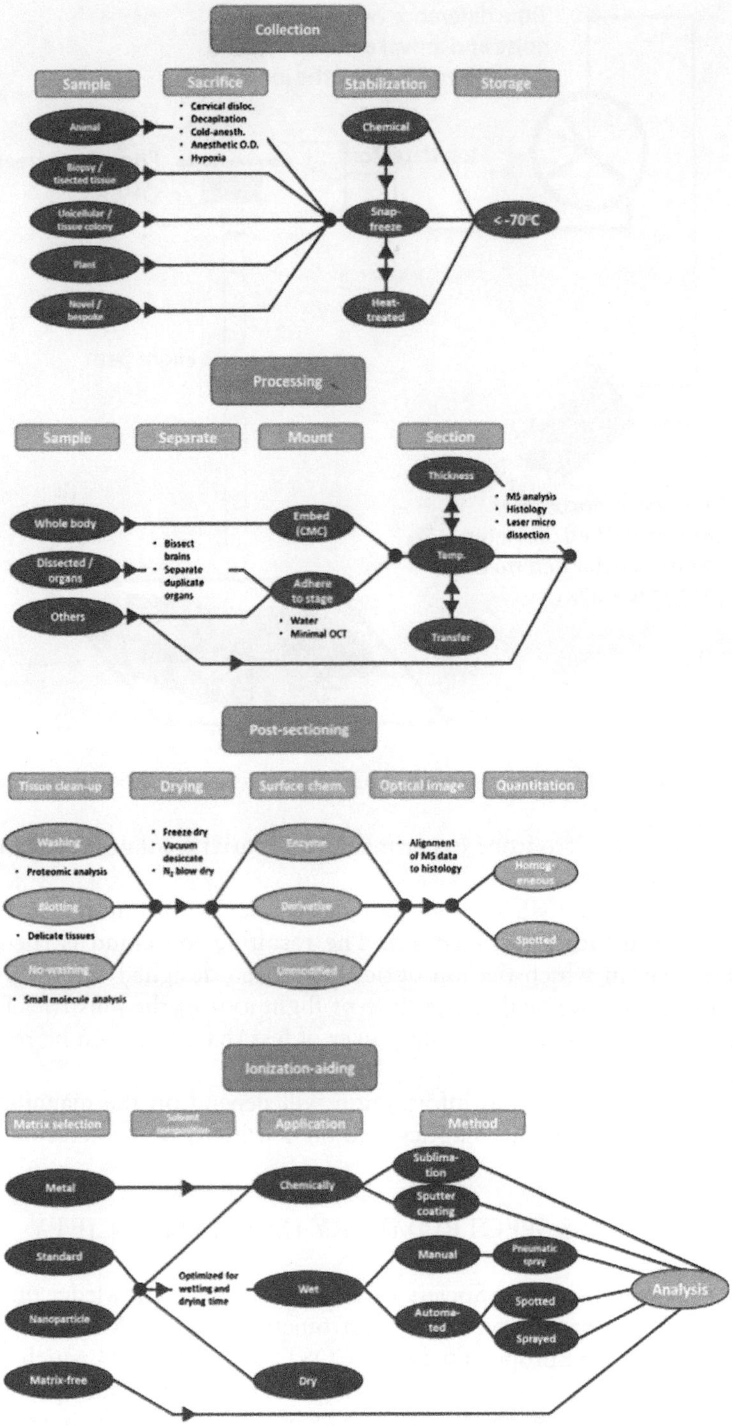

FIGURE 1.1 Schematic of the major stages in sample preparation for MS imaging broken down into the four major components, encompassing collection, processing, post-sectioning, and ionization-aiding treatments.

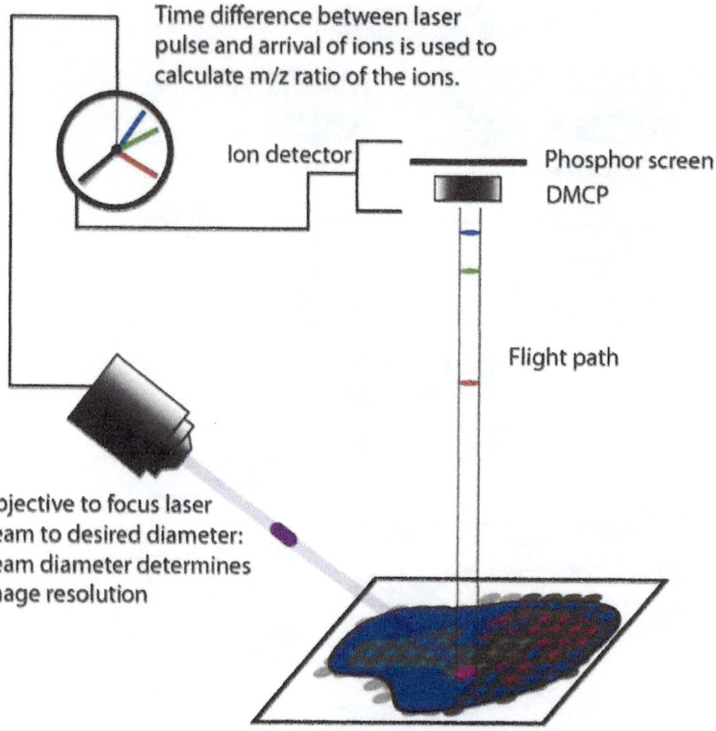

FIGURE 1.2 Schematic overview of microprobe imaging mass spectrometry setup.

In a conventional microprobe experiment, the spatial resolution is restricted by the spot-size of the laser (Figure 1.2).

In microscope mode IMS a large-diameter (typically 200 µm) laser beam is used as a desorption and ionization medium. The resulting ion cloud is then transferred by the TOF system in which the ion optics have been designed to retain the original position of the ions orthogonal to the time-of-flight axis as the ions travel through the mass spectrometer. An image resolving power of less than 4 µm can be reached by this methodology.

The resolution of the spatial information will depend on the magnification of the microscope, the quality of the ions optics, and the sensitivity of the detector (Figure 1.3).

1.4 MASS SPECTROMETRY IMAGING SOCIETY

The Mass Spectrometry Imaging Society (5) aims to share the knowledge on MS imaging.

A community of practice on mass spectrometry imaging (MSI) emerged in 2012 under the auspices of the European network COST Action BM1104 which aimed at connecting scientists working in the rapidly growing field of MSI. The European nature of the COST funding has meant these early activities (conferences, workshops, training courses) had a European focus, but they were open to worldwide participation. Consequently, the community had over 200 active members from all continents.

It was decided in October 2015 to formally establish a Mass Spectrometry Imaging Society to continue the activities of the COST network and to include other activities

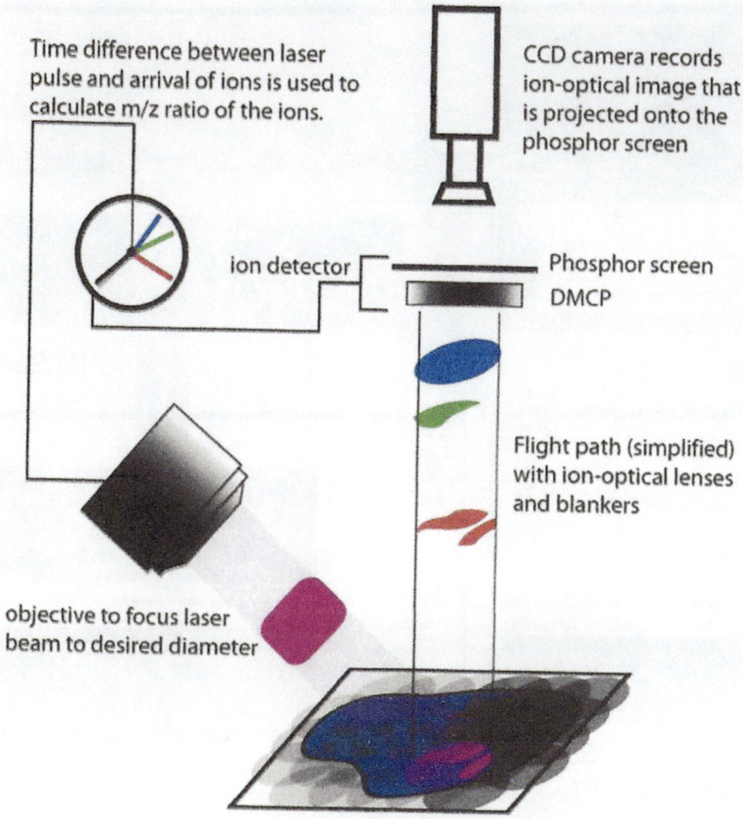

FIGURE 1.3 Schematic overview of microscope imaging mass spectrometry setup.

and all practitioners in the field of MS imaging. The possibility of joining an existing society, e.g., to form a focus group within the International Mass Spectrometry Foundation (IMSF) was evaluated and discussed during 2016. As a result of extensive discussions during OurCon IV (Ustron, Poland, October 2016) it was almost unanimously decided to form the Mass Spectrometry Imaging Society, with the legal entity being in Switzerland.

As a result, the society was founded on February 10, 2017 in Basel, Switzerland.

On the link *Software* users are able to download software. One can make a choice between BioMap, MMSIT, 4000 Series Imaging, AnalyzeThis!, Axima2Analyze, MSImageView, and msIQuant.

One page of this website is dedicated to imzML. This page *imzML* is intended to provide all information necessary to implement imzML (Figure 1.4).

1.5 imzML

imzML is a common data format for MS imaging (6, 7). The purpose of imzML is to facilitate the exchange and processing of mass spectrometry imaging data. Several imaging MS software tools support it. The advantage of this format is the flexibility to exchange data between different instruments and data analysis software.

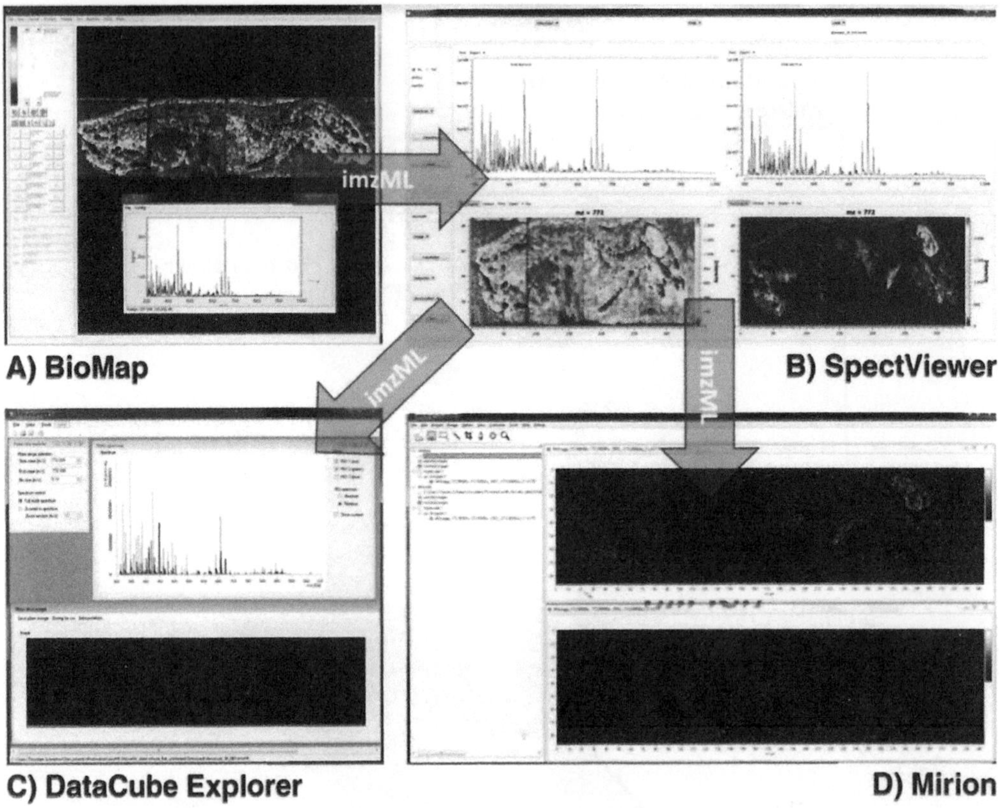

FIGURE 1.4 Example of sequential processing of an MS imaging data set based on imzML.

imzML was developed in the framework of the EU-funded project COMPUTIS (8). The main goals during the development were:

- Complete description of MS imaging experiments
- Efficient storage of (very large) data sets

imzML is not limited to MS imaging, but is also useful for other MS applications generating large data sets such as LC-FTMS.

The data format imzML was developed to allow the flexible and efficient exchange of MS imaging data between different instruments and data analysis software. For this purpose, the MS imaging data are divided in two separate files. The mass spectral data are stored in a binary file to ensure efficient storage. All metadata (e.g., instrumental parameters, sample details) are stored in an XML file which is based on the standard data format mzML developed by HUPO-PSI (Human Proteome Organisation-Proteomics Standards Initiative). See also www.psidev.info/ (9). The original mzML controlled vocabulary was extended to include specific parameters of imaging mass spectrometry (such as (x,y) position and spatial resolution). The two files (XML and binary) are connected by offset values in the XML file and are unambiguously linked by a universally unique identifier. The resulting data sets are comparable in size to the raw data, and the separate metadata file allows flexible handling of large data sets. Several imaging MS software tools already support imzML. This allows a choice from a (growing) number of processing tools. One is no longer limited to proprietary software, but is able to use the processing software which is best suited for a specific question or application. On the other hand, measurements from different instruments can be compared within one software application using identical settings for data processing. All necessary information for evaluating and implementing imzML can be found at http://imzml.org/wp/imzml/ (10).

1.6 DATA ANALYSIS SOFTWARE

There are many free software packages available for the visualization and mining of imaging mass spectrometry data. Converters from the Thermo Fisher format, Analyze format, GRD format, and Bruker format to the imzML format were developed by the Computis project (8). Some software modules are also available for viewing mass spectrometry images in the imzML format: Biomap (Novartis, free) (10), Datacube Explorer (AMOLF, free) (11) EasyMSI (12), Mirion (Justus-Liebig Universität) (13), MSiReader (North Carolina State University, free) (14), and SpectralAnalysis (15).

For processing .imzML files with the free statistical and graphics language R, a collection of R scripts is available, which permits parallel-processing of large files on a local computer, a remote cluster, or on the cloud.

Another free statistical package for processing imzML and Analyze 7.5 data in R exists, Cardinal (16).

SPUTNIK (17) is an R package containing various filters to remove peaks characterized by an uncorrelated spatial distribution with the sample location.

1.7 MALDI – SIMS – DESI – LESA – LA-ICP

MALDI is an ionization technique using laser energy. In a matrix ions are created from large molecules with minimal fragmentation. It is comprised of photochemical (ionization and fragmentation) and mechanical (desorption) events. In MALDI, analytes are incorporated into organic matrices in a solid or viscous phase (ionic liquid matrices). A pulsed laser irradiates the sample to facilitate the ionization/desorption reactions. During laser excitation, matrices take up and redistribute the laser energy for the desorption and ionization of analytes.

In SIMS the composition of solid surfaces and thin films is analyzed by sputtering the surface of the sample with a focused primary ion beam and collecting and analyzing ejected secondary ions. The system is maintained under an ultrahigh vacuum to prevent atmospheric interference with the primary and secondary ions.

When a solid sample is sputtered by primary ions a fraction of the particles emitted from the target is ionized. SIMS consists of analyzing these secondary ions with a mass spectrometer.

DESI ionizes compounds with little sample preparation and without a vacuum system, and it involves spraying electrically charged solvent droplets onto a sample surface, accompanied by a high-velocity gas jet. The generally accepted model for DESI analyte ionization involves analyte dissolving in a thin layer of the DESI solvent before being ejected in secondary droplets by impact from the primary spray. The charged secondary droplets evolve into gas-phase ions similar to the theorized electrospray mechanism due to electrostatic and pneumatic forces. The desorbed gas-phase ions are then sucked into a mass spectrometer for detection through a custom-built electrically charged metal inlet.

Liquid extraction surface analysis (LESA) is a high-throughput tool for ambient mass spectrometry. A solvent droplet is deposited from a pipette tip onto a surface and maintains contact with both the surface and the pipette tip for a few seconds before being re-aspirated. The technique is particularly suited to the analysis of trace materials on surfaces due to its high sensitivity and low volume of sample removal.

Laser ablation inductively coupled plasma mass spectrometry (LA-ICP-MS) begins with a laser beam focused on the sample surface to generate fine particles. This process is known as laser ablation. It is a process in which a laser beam is focused on a sample surface to remove material from the irradiated zone. Laser ablation is a dynamic technology platform for performing chemical analysis. LA also creates a flux of fine particulates that can be readily transported to a secondary source for ICP-MS analysis.

The ablated particles are then transported to the secondary excitation source of the ICP-MS instrument for the digestion and ionization of the sampled mass. The excited ions in the plasma torch are subsequently introduced to a mass spectrometer detector for both elemental and isotopic analysis.

For more information on these MSI techniques and their applications in foodstuffs see the following chapters. Other, less used techniques are commented on in the last chapter of this book.

REFERENCES

1. Buchberger A.R., DeLanev K., Johnson J., Li L. Mass spectrometry imaging: A review of emerging advancements and future insights. *Analytical Chemistry*, 2018, 90 (1), 240–265.
2. Römpp A., Spengler B. Mass spectrometry imaging with high resolution in mass and space. *Histochemistry and Cell Biology*, 2013, 139 (6), 759–783.
3. Goodwin R.J.A. Sample preparation for mass spectrometry imaging: Small mistakes can lead to big consequences. *Journal of Proteomics*, 2012, 75, 4893–4911.
4. Klerk A.L., Altelaar A.F.M., Froesch M., McDonnell L.A., Heeren R.M.A. Fast and automated large-area imaging MALDI mass spectrometry in microprobe and microscope mode. *International Journal of Mass Spectrometry*, 2009, 285, 19–25.
5. https://ms-imaging.org/wp/mass-spectrometry-imaging-society/
6. Römpp A. et al. imzML: Imaging mass spectrometry markup language: A common data format for mass spectrometry imaging. In: Hamacher M., Eisenacher M., Stephan C. (eds), *Data Mining in Proteomics. Methods in Molecular Biology (Methods and Protocols)*, vol. 696. Humana Press, 2011.
7. Schramm T., Hester Z., Klinkert I., Both J.-P., Heeren R.M.A, Brunelle A., Laprévote O., Desbenoit N., Robbe M.F., Stoeck M., Spengler B., Römpp A. imzML — A common data format for the flexible exchange and processing of mass spectrometry imaging data. *Journal of Proteomics*, 2012, 75 (16), 5106–5110.
8. http://www.computis.org
9. http://www.psidev.info/
10. http://imzml.org/wp/imzml/
11. https://amolf.nl/download/datacubeexplorer
12. https://www.easy-msi.com/

13. Paschke C., Leisner A., Hester A., Maass K., Guenther S., Bouschen W., Spengler B. Mirion--a software package for automatic processing of mass spectrometric images. *Journal of the American Society for Mass Spectrometry*, 2013, 24 (8), 1296–1306.
14. https://msireader.wordpress.ncsu.edu/
15. https://github.com/AlanRace/SpectralAnalysis
16. https://cardinalmsi.org/
17. https://cran.r-project.org/web/packages/SPUTNIK/SPUTNIK.pdf

CHAPTER 2

Imaging Mass Spectrometry for Small Molecules

Fangbiao Li

CONTENTS

2.1 INTRODUCTION

Mass spectrometry-based bioanalytical methods play an important role in the entire process of searching for new medicines to improve the quality of human life. High-performance liquid chromatography coupled to a tandem mass spectrometer (HPLC–MS/MS) has become the standard tool for the determination of pharmaceuticals in various *in vitro* and *in vivo* samples stemming from drug discovery and development experiments [1–5]. However, the major disadvantage of the HPLC–MS/MS technique is that it does not readily provide information on the distribution of the administrated drug and its metabolites.

Alternatively, imaging mass spectrometry (IMS) using matrix-assisted laser desorption/ionization (MALDI) can be used to simultaneously map the distribution of pharmaceuticals in thin tissue sections to determine how a drug is distributed in animal tissues [6–9]. MALDI–IMS has been extensively employed to measure macromolecules such as peptides and proteins in tissue sections [10–13] (Figure 2.1). Although MALDI–IMS has been applied almost exclusively as an analytical tool for biopolymers, the potential of using this technique for both the qualitative and quantitative determination of small molecules (<1500 Da) such as enzyme subtracts [14], acetyl-choline, 3,4-dihydroxy-phenylalanine [15], vitamins [16], and others [17, 18] has been demonstrated. Several teams of researchers have extended MALDI–MS imaging methodology to visualize the

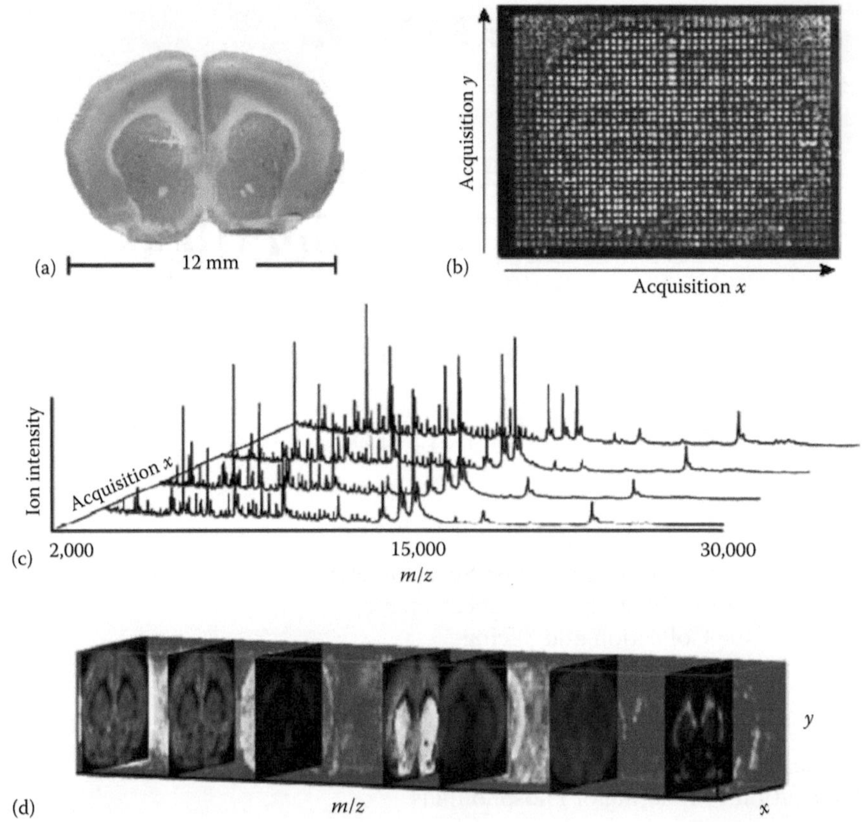

FIGURE 2.1 General overview of MALDI–IMS. (a) Fresh section cut from sample tissue. (b) Mounted section after matrix application using a robotic picoliter volume spotter. (c) Partial series of mass spectra collected along one row of coordinates (x-axis). (d) Three-dimensional volumetric plot of complete dataset with selected m/z "slices" or ion images. Principal axes are x, y, and m/z. Color of each voxel is determined by ion intensity. (From Cornett, D. et al., *Nat. Methods*, 4, 828, 2007. With permission.)

distribution of candidate drug compounds in different regions of biological tissue sections [19–24]. In this chapter, we describe the use of MALDI mass spectrometric imaging as a tool for the pharmaceutical analysis of drug compounds and their metabolites in the drug discovery and development process. This technique offers significant advantages over alternative technologies, such as autoradiography. For example, MALDI–MS imaging can distinguish intact drugs from their metabolites without the addition of an isotope label. This allows distribution studies of lead compounds to occur much earlier in the drug-discovery process and allows compounds with unfavorable distribution characteristics to be rapidly discarded. In addition, the reliance on radioactive isotopes can be reduced, which is advantageous from both an economical and environmental standpoint.

Distribution studies of drug candidates in animals are crucial in both drug discovery and development. These studies provide information about where the drug accumulates in the body. For example, it may be important to know if the test compound accumulates in the target organ or if it selectively accumulates in other organs or tissues (such as the brain, which could result in neurotoxicity or other unwanted side effects). It is also important to know how long a drug remains in the body, and MALDI–IMS can help to measure that parameter as well. The accurate assessment of preclinical absorption/distribution/metabolism/excretion (ADME) and toxicological parameters of new chemical entities (NCEs) is essential for deciding whether or not to develop the NCE. Techniques currently in use for assessing the distribution of drug candidates in tissues will be discussed; these techniques include positron emission tomography (PET), magnetic resonance imaging (MRI), autoradiography, and secondary ion mass spectrometry (SIMS). Their strengths and limitations will be addressed, especially with regard to the spatial analysis of drugs and metabolites in tissues. The technique of MALDI–IMS will be discussed, and the examples of its use will be presented. Finally, the further development of the technique will be addressed, focusing on improvements and advances necessary to allow the technology to reach its full potential.

2.2 CURRENT IMAGING TECHNIQUES

Several analytical techniques exist to image the distribution of small molecules in the body. Generally, they can be divided into two groups: noninvasive *in vivo* techniques and *ex vivo*/*in vitro* techniques. *In vivo* techniques, such as PET [25–32] and MRI [32–36], allow the distribution of a drug to be evaluated over time in a living animal. In contrast, the *ex vivo*/*in vitro* techniques, such as autoradiography [29, 30], require the removal of the tissue of interest and thus can only image the distribution of a drug at a fixed time.

All these techniques require labeled drugs. The visualization of PET requires that a drug contain a radioactive positron emitter, such as ^{11}C, ^{13}N, ^{15}O, or ^{18}F [32]. Similarly, a radioactive isotope, typically ^{14}C, ^{3}H, or ^{125}I, is necessary for a compound to be visualized with autoradiography. While MRI does not require a radioactive isotope, only isotopes with nuclear spin, such as ^{1}H, ^{13}C, and ^{19}F, are visible with MRI [32, 33]. Because the sensitivity of MRI depends on the magnetic properties of the monitored nucleus and its natural abundance, drugs typically need to contain ^{19}F or be enriched in ^{13}C in order to be effectively monitored in the body. Contrast agents, which include paramagnetic atoms such as gadolinium, iron, and manganese, are often used to increase the contrast in MRI images and are frequently used as models for drugs that cannot be imaged on their own [33, 34]. The main limitation with using a label or tracer for imaging, especially in terms of drug metabolism, is that only the label and not the parent compound is imaged. Thus, the experimentally determined distribution of a drug in a tissue may be inaccurate if extensive metabolism has occurred.

In vivo imaging techniques have several advantages for drug-discovery applications, including the ability to examine a drug's actions directly in humans and the ability to re-measure the same individual under different conditions. However, there are significant limitations. The spatial resolution is limited with PET, and MRI–PET has a resolution of ~3–6 mm [26, 32, 33], while MRI has a resolution of ~2 mm [34, 35], but can be improved to 700–800 μm [36] with some modifications to the instrument used. This limits the size regime in the body in which meaningful distribution data can be acquired. The use of a labeled compound typically occurs late in the drug development process due to the increased cost and time involved to synthesize the labeled compound. Additionally, there are time constraints with PET imaging due to the half-lives of the positron emitters. The half-lives of some commonly used positron emitters are ^{15}O, 2 min; ^{13}N, 10 min; ^{11}C, 20 min; and ^{18}F, 110 min. Thus a drug must be synthesized, administered, and monitored on the time-scale of the half-life of the radiotracer.

Autoradiography is one of the most common techniques used today in the pharmaceutical industry for examining the distribution of a drug candidate in the body *ex vivo* [37, 38]. In quantitative whole-body autoradiography (QWBA), a radiolabeled drug is administered to an animal and, after a specific time, the animal is sacrificed, flash-frozen, and sectioned. The radioactivity in the sections is then analyzed, and the result is an image showing where the drug has accumulated in various organs. Individual organs may be dissected and analyzed separately as well. This technique is also widely used in psychopharmacology for *in vitro* experiments, in which a radiolabeled compound is incubated with a section of brain tissue, and the resulting image is used to determine where in the brain the compound is binding in order to locate the specific receptors of the compound [38, 39].

Autoradiography has significant advantages over the noninvasive imaging techniques, especially in terms of drug discovery. As it is typically performed on laboratory animals, it occurs earlier in the drug-discovery process. Additionally, the resolution is better that PET and MRI, ranging from a few micrometers for film to ~30–50 μm for the β-imager [38, 39], ~70 mm for microchannel plates [40, 41], and ~25–100 μm for commercial phosphor-imaging systems (Fujifilm). In contrast, the time required to acquire an autoradiographic image varies substantially and can be prohibitively long. It often takes weeks or months to acquire an image on film, while the same image may be acquired on the β-imager in 8–12 h [38, 42]. Also, because a radiolabel is required, the extra cost and time required for additional synthesis are still limitations.

2.3 PRINCIPLES OF MALDI IMAGING MASS SPECTROMETRY

Due to its molecular specificity, there is a great deal of interest in using mass spectrometry as an imaging tool. Instruments used for IMS can be classified according to how ions are generated from the sample either by irradiation by pulsed laser or bombardment by energetic particles [43]. SIMS systems use particle bombardment with a continuous beam of highly focused, energetic ions such as Cs^+, Au_3^+, and C_{60}^+. Laser-based systems include MALDI and laser desorption ionization (LDI) instruments.

SIMS has been used for imaging surfaces for more than 40 years [44]. The primary ion beam has an energy of ~25 keV which desorbs ions from the surface of the sample. These secondary ions are predominantly elements, atomic clusters, and organic fragments that are typically analyzed by time-of-flight (TOF), quadrupole, or magnetic sector instruments. While protonated molecules can be formed, for example, several analyses have been reported involving the $[M + H]^+$ ions of cholesterol [45], benzodiazepines [46], and crystal violet [47], the secondary ions formed most abundantly include elemental ions (e.g., Na^+, K^+) and molecular fragment ions. Thus, its primary use has been for the localization of inorganic and atomic species [44].

Applications of SIMS imaging to biological/pharmaceutical analyses have been in two key areas: atomic imaging of drugs [48–50] and tissue mapping via molecular imaging of fragment ions [51–53]. In the first application, the drug of interest must be effectively labeled; it must contain an atom not natively found in cells or tissues, and only that atom is monitored. The other main application of SIMS to biological imaging has focused on tissue mapping. Compared to the traditional imaging techniques discussed earlier, the benefits of the SIMS imaging techniques include excellent spatial resolution (submicrometer to ~1 μm) and high sensitivity and molecular specificity. One of the main drawbacks, however, especially in regards to metabolite localization, is the low efficiency of molecular ionization [47]. Thus it is difficult to differentiate a compound and structurally similar metabolites if they share the same label or if they form the same fragment, because that is what is typically analyzed. While advances are being made toward increasing molecular ion yields, especially utilizing cluster ion beams such as C_{60}^+ and SF_5^+ [47, 54], the full potential of this technology for pharmaceutical imaging has not been reached.

MALDI–IMS employs a matrix solution to mix with the analytes and to form cocrystals. These crystals absorb energy at the wavelength of the laser beam resulting in desorption and ionization of the analytes that were included in the crystals. The concept of MALDI–IMS was introduced in 1997 by Caprioli and coworkers [55] for rapid and direct profiling of the analytes within a tissue section or an organ. In this process, a matrix is uniformly deposited over various tissue sections to extract analytes into the matrix and to produce matrix crystals (Figure 2.2) [56]. A raster of the organ or whole-body sections containing the compound of interest under a stationary laser beam is then performed over a predetermined two-dimensional array to generate ion plumes directly from the tissue sections in a MALDI plate array as depicted in Figure 2.3 [56]. For the small-molecule application, commercially available MALDI source instruments have been utilized. These instruments utilize a large sample plate sufficient to contain whole tissue slices. The movement of the sample stages is automatically accomplished in the *x*- and *y*-directions to locate the edges of the tissue sample and to define the exact region of interest (Figure 2.2).

During an imaging mass spectrometric experiment, MALDI–MS signals of small molecules from tissue sections within a user-defined area are first obtained as a function of acquisition times which are associated with the location of an array of pixels as given in Figure 2.4 [56]. Thus, two-dimensional ion maps of biological tissues are reconstructed with drug signals of a given *m/z* value monitored in each spectrum from each pixel within a defined array to provide specific molecular images as demonstrated in Figure 2.5 [56]. In order to detect small molecules in complex biological tissue sections

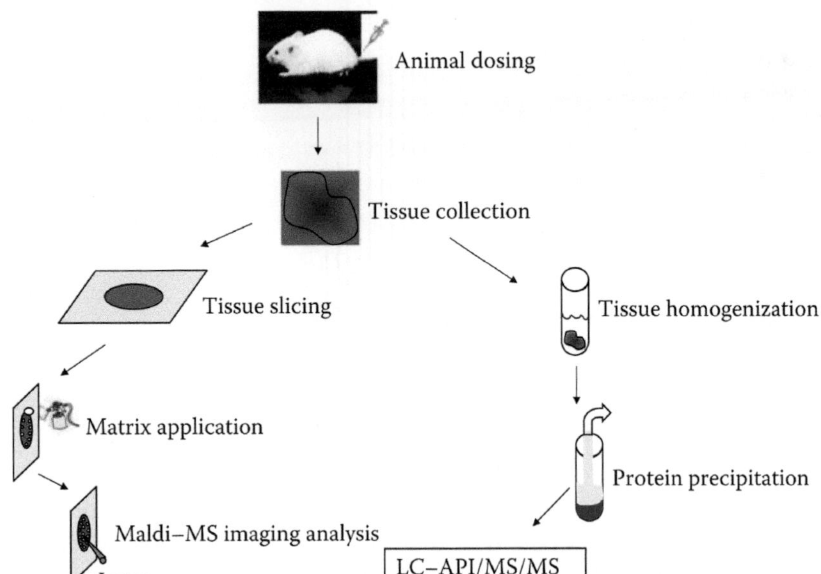

FIGURE 2.2 General procedure for MALDI–MS imaging analysis and conventional quantitative HPLC–MS analysis.

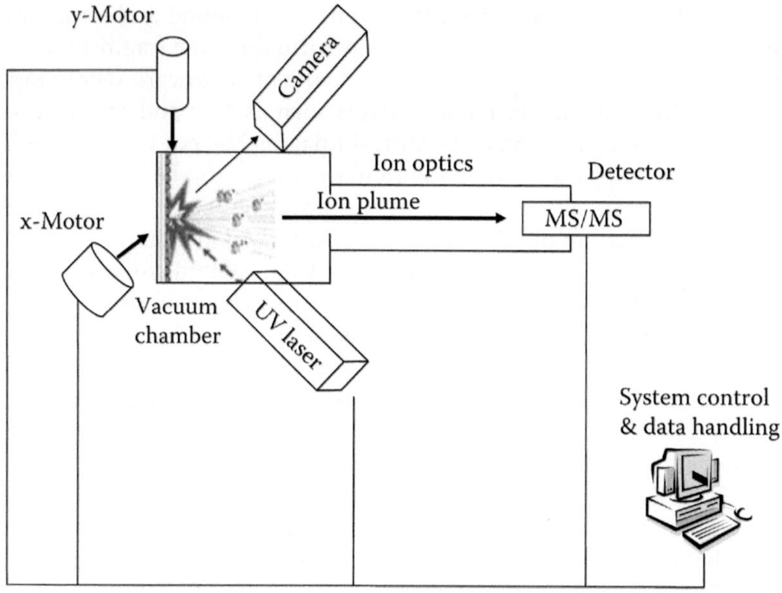

FIGURE 2.3 Diagram for an automated MALDI–IMS system used for direct analysis of pharmaceuticals in tissues. (From Hsieh, Y. et al., *J. Pharmacol. Toxicol. Methods.*, 55, 193, 2007. With permission.)

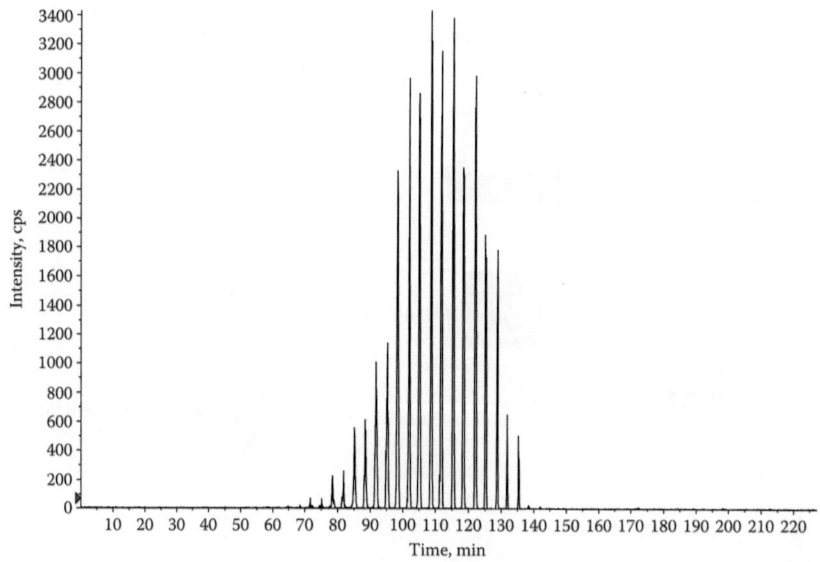

FIGURE 2.4 MALDI–MS signals from a droplet deposited in the rat brain tissue section containing a drug candidate are plotted against acquiring time which is directly related to the location of each pixel. (From Hsieh, Y. et al., *J. Pharmacol. Toxicol. Methods.*, 55, 193, 2007. With permission.)

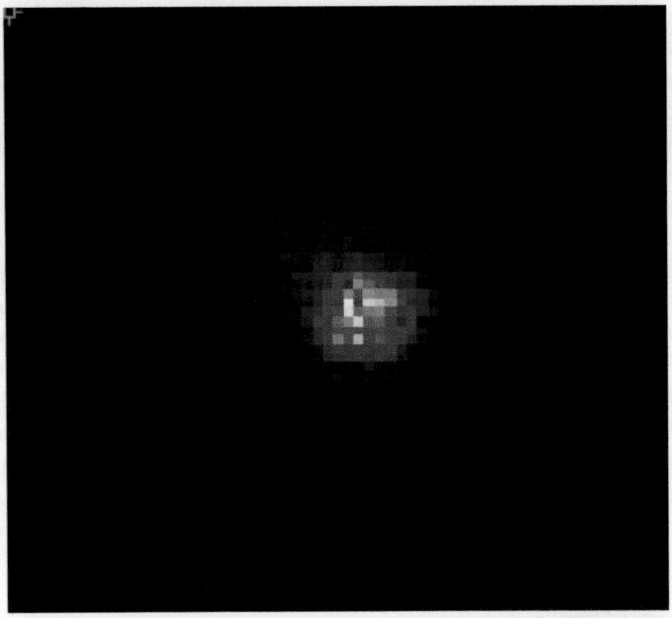

FIGURE 2.5 An ion image of a spot in the rat brain slice was obtained after reconstruction of MALDI–MS signals derived from the spot of chemical residue while “rastering” the tissue sample under the laser beam. (From Hsieh, Y. et al., *J. Pharmacol. Toxicol. Methods*., 55, 193, 2007. With permission.)

using MALDI, a tandem mass spectrometer is required in order to separate the analyte ions from the background interference ions produced by the matrix [22]. In most MS/MS systems, two analyzers are connected together by a collision cell. A protonated molecule is generated in the ion source. Then the first analyzer selects the ion as the precursor ion for collisionally activated dissociation (CAD). This precursor ion passes through a collision cell, which breaks the precursor ion into product ions. Analysis of the resulting product ions by the second analyzer generates a mass spectrum of the fragment ions, providing structure information. By monitoring the transition of a selected precursor ion to its dissociated product ions, tandem mass spectrometry systems give researchers a tool for distinguishing between precursor ions of similar mass-to-charge ratios, such as matrix ions and small molecule drug candidate compounds [22]. For example, a hybrid quadrupole-time-of-flight (QqTOF) analyzer was utilized in order to resolve some of the mass spectrometric interference from the matrix, as CAD fragmentation of protonated drug molecules could be performed [19, 21, 22].

2.4 FUNDAMENTALS OF IMAGING MASS SPECTROMETRY

2.4.1 Tissue Collection and Slicing

Tissue collection and slicing are critical to the spatial integrity of measured molecular distribution [57]. Inappropriately handling tissue samples in the sample preparation steps may cause delocalization and degradation of the analytes. A typical study may involve samples collected over a lengthy period of time, and standardized procedures are therefore required to minimize experimental variability over the time course of the study. Ideally, animal tissues are frozen immediately following collection to prevent drug dispersing in tissue. For the analysis, animal tissues are sliced to expose the area of interest. Cryostats are normally used as a standard tool for slicing frozen tissue to reducing sample contamination [56]. Contamination of the tissue surface due to the use of optimal cutting temperature (OCT) polymer as an embedding medium for stabilizing the organ specimens should be avoided because it would lead to ionization suppression in MALDI–IMS analysis [13, 58]. Recently, it has been reported that distilled water can be used as an embedding medium to overcome this problem [59]. There are several direct ways to transfer tissue slices to the sample plate [56]. First, the frozen tissue slice is gently positioned on the cold target plate (approximately –15°C), using an artist's brush, and thaw-mounted onto the plate by quickly moving out of the cryostat chamber. Second, the frozen section adheres to the MALDI plate held at room temperature. Third, the section is stuck to a double-sided transparent tape which is then attached to the sample plate. In addition, blotting the tissue onto cellulose or carbon-conducting membranes is an indirect way to transfer the sections. Tissue containing drugs subject to photodegradation should be maintained in a dark container as much as possible during the processing steps.

2.4.2 MALDI Matrix Application

Once the section is mounted to the sample plate with the desired orientation, matrix solution is deposited on the tissue surface by electrospray deposition, aero spray, or using robotics to deposit small matrix droplets across the tissue surface before MALDI analysis [13]. The most common and least expensive devices available for applying matrix are hand-held aerosol sprayers or air brushes. The main advantage of these devices is that, with careful application, a dispersion of very small droplets of matrix solution can be delivered onto the sample surface to form a homogeneous layer of crystals. An alternative to aerosols is the delivery of matrix solution to the sample in the form of very small droplets from precision ejectors. Two different types of ejectors have been incorporated into such devices: inkjet-style piezo nozzles [60] and focused acoustic dispensers [61]. Both designs operate by drop-on-demand, producing highly reproducible drops of approximately 1000 pL that dry to form crystalline matrix spots of ~150 μm diameter. Aerosol-generated droplets can be smaller in size, but spotting devices offer advantages of droplet uniformity and precision of placement, both of which translate into greater control over analyte extraction. High- resolution images with MALDI–IMS necessitate a uniform coating by the matrix, no redistribution of surface analytes, and linearity of MALDI signals after matrix application. In order to produce a homogeneous crystal layer over the tissue surface, spraying multiple coats of matrix across the surface of the tissue is recommended. Typically, one coating cycle consists of passing the sprayer two to three times across the surface of the tissue and allowing the tissue to dry between cycles. The sample plate is held vertically about 20–30 cm from the sprayer nozzle. The sample is allowed to dry for at least 2–5 min before the next coating cycle to avoid the deposition of a large quantity of solvent in any one region of the tissue. The combination of the spray rate and spray distance should be adjusted to avoid excessive wetting of the tissue which could lead to analyte dispersion. This process iterates usually more than ten cycles and serves to redissolve and recrystallize the matrix, enhancing the incorporation efficiency of analytes in the crystals. Contaminants such as salts and phospholipids in tissues may significantly inhibit cocrystallization, resulting in low MALDI intensities of the analytes. Some researchers have proposed washing samples with weak organic solvents such as 70% ethanol prior to matrix application to eliminate endogenous materials from the native tissue and to improve the crystallization process and the ionization efficiency for biopolymers [57]. However, this is not recommended for small-molecule applications because there is the potential of altering the spatial integrity of drug-related components on tissues. Washing of tissue samples with water has been reported recently to improve signals and reduce the background noise in some circumstances [62].

2.4.3 Experimental Conditions

The selection of matrix and matrix solution conditions such as solvent compositions, pH, and the rates of cocrystal growth can affect the quality of mass spectra for both macromolecules and small molecules [63]. The success of the MALDI–IMS applications to an analyte in tissue is strongly dependent on the choice of appropriate matrix materials. The common UV-absorbing molecules (Figure 2.6) used as matrices for MALDI analysis are benzoic acid–based components with low molecular weights (<500 Da) which dominate the low-mass range background for a typical MALDI–MS spectrum, further challenging the advancement of MALDI for the analysis of small molecules.

Because of the nature of biological tissues, the growth of crystal is more complicated on tissue than on an inert plate where a small volume of matrix is mixed with a neat drug solution. Strong acids such as 0.1% trifluoroacetic acid (TFA) that are normally added to the matrix solution to assist the protonation of proteins have been found to have a marginal effect on the ionization efficiency for small molecules [56]. For small molecules, a higher matrix concentration tends to produce better quality of mass spectra. The solvent composition may have a substantial effect on the responses of the targeted compounds dependent on their effectiveness of analyte extraction and crystal formation. Laser wavelength, fluence, and spot size on the sample have a strong impact on the desorption/ionization process [64]. The MALDI–MS signals of drug-related compounds from tissues were detected when the fluence values were higher than the fluence threshold which is normally greater than those from the neat solution. The relation between laser radiation power and the MALDI responses from the tissue surface is dependent on the compounds of interest. For clozapine, the MALDI–MS signal amplitude improved with an increase in laser beam exposure up to around 70 μj and declined steadily with higher laser power (Figure 2.7) [21]. For norclozapine, the ion production was roughly proportional within an increasing laser exposure range under the same experimental conditions.

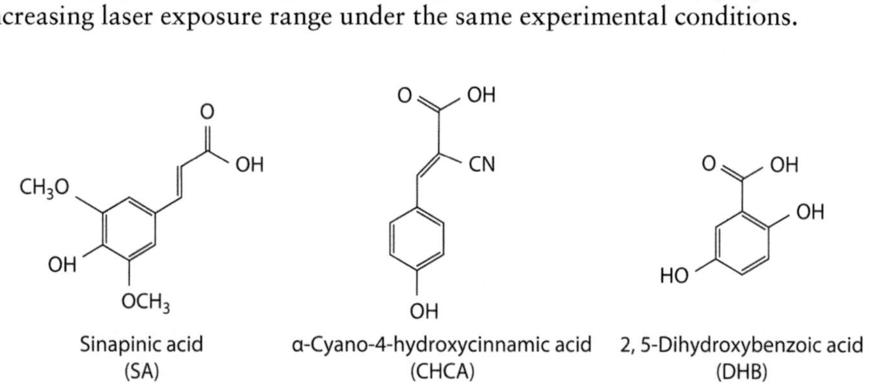

FIGURE 2.6 Chemical structures of three common MALDI matrices.

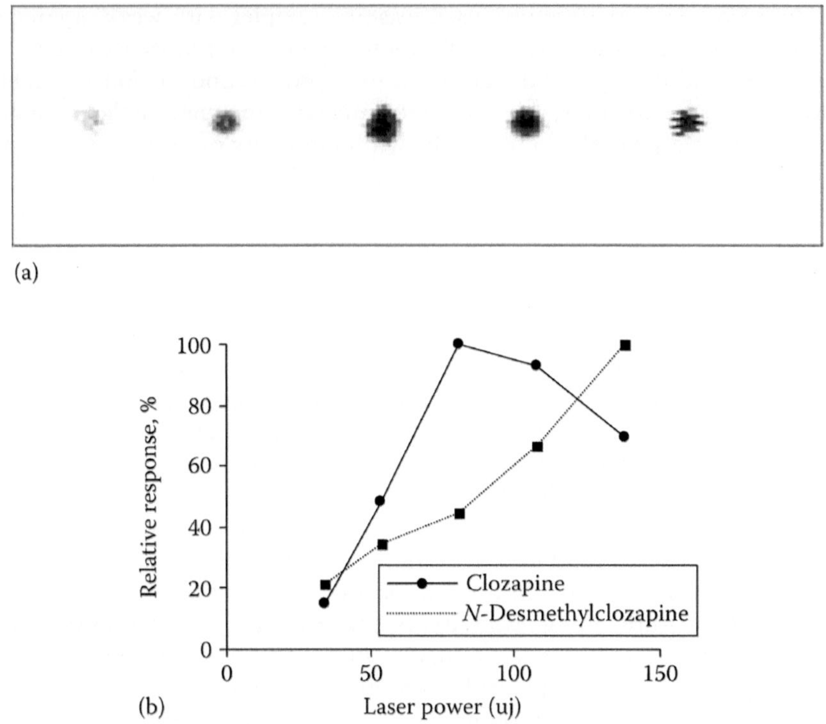

FIGURE 2.7 (a) MALDI–MS images of clozapine in rat brain slice as a function of laser fluence. (b) Relative MALDI responses of clozapine and norclozapine in rat brain slice as a function of laser fluence. (From Hsieh, Y. et al., *Rapid Commun. Mass Spectrom.*, 20, 965, 2006. With permission.)

MALDI imaging resolution is governed by both crystal size and laser diameter. Generally smaller crystal sizes yield better imaging spatial resolution. With crystal diameters smaller than the laser beam, typically 50–200 μm depending on the instrument, imaging resolution is generally limited to the laser diameter [57]. Acquisition times for tissue imaging rely on several instrumental parameters such as spatial resolution requirements, the laser repetition rate, spot-to-spot sample repositioning transfer time, and data processing. In order to obtain higher resolution (more pixels on a given tissue slice), one needs to use longer acquisition times. With current instrumentation, it can take several hours to obtain the data for one tissue sample. As lasers with faster repetition rates and improved electronics become available, one could reduce the acquisition times from hours to minutes [10].

2.4.4 Imaging Drugs in Tissue Sections by MALDI

MALDI signals were found to be proportional to the densities of pharmaceuticals in tissues, although different regions within the same organ section and different types of tissues might demonstrate different surface properties (Figure 2.8) [21].

The linearity of analyte responses and the ionization suppression degree due to the heterogeneity within a tissue section was addressed by making a calibration curve for

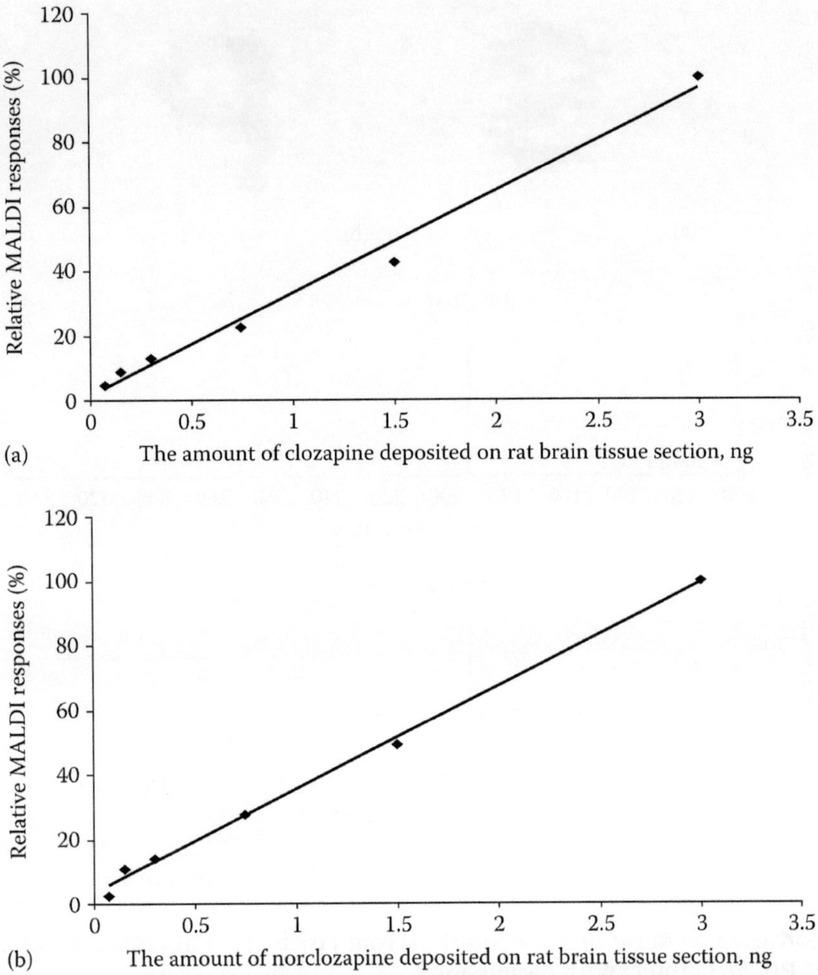

FIGURE 2.8 Linear MALDI responses of (a) clozapine and (b) norclozapine with increasing concentrations from six individual droplets spotted in rat brain tissue slice. (From Hsieh, Y. et al., *Rapid Commun. Mass Spectrom.*, 20, 965, 2006. With permission.)

the semiquantitation of a parent drug and its metabolite by depositing several droplets with increasing concentrations on a blank tissue section. Good linearity data suggested that the impact of ionization suppression from one location to another location within the tissue section is not significant [19, 21]. A consistent MALDI signal from the matrix outside the tissue section indicated a uniform tissue coating by the matrix. Another useful means to evaluate the feasibility of MALDI–IMS technique was to map the location of the spiked analytes within a blank tissue section. For example, the letters S and P were inscribed by connecting microdroplets containing clozapine and norclozapine through a microsyringe, respectively (Figure 2.9) [21]. The ion transitions from *m/z* 327 to *m/z* 192 and from *m/z* 313 to *m/z* 192 for clozapine and norclozapine, respectively, were monitored at discrete spots over the entire rat brain section. The MALDI–IMS images of clozapine and norclozapine clearly indicated these two letters reflected on the tissue sections without distortion. These experiments suggested that the redistribution of surface

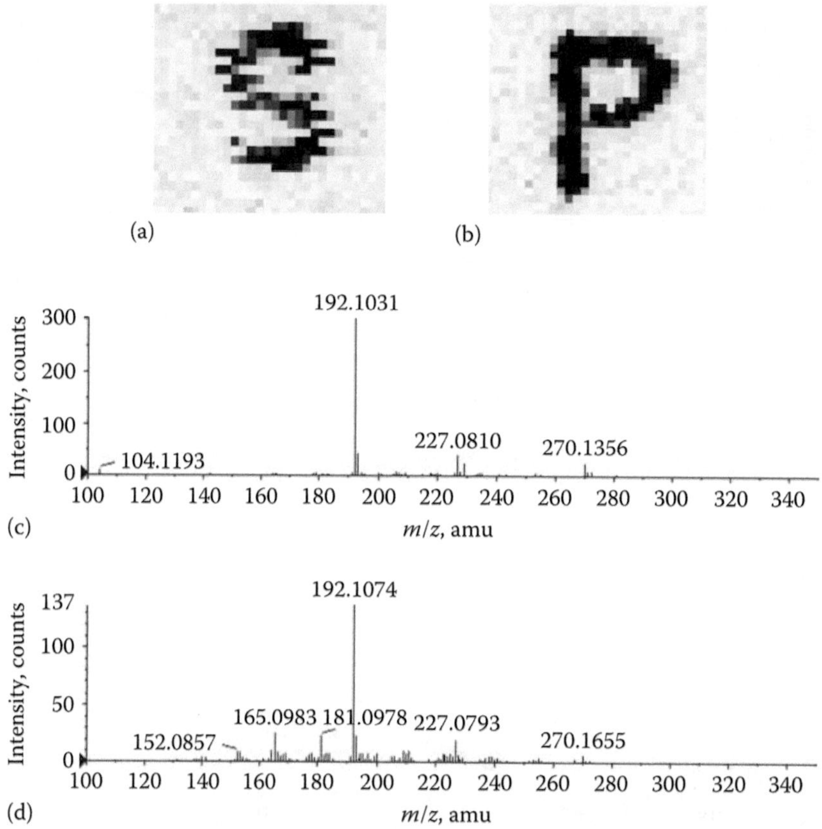

FIGURE 2.9 MALDI–MS images of letters (a) S and (b) P containing norclozapine and clozapine in rat brain sections, respectively. MALDI–MS/MS spectra of (c) clozapine and (d) norclozapine in rat brain sections. (From Hsieh, Y. et al., *Rapid Commun. Mass Spectrom.*, 20, 965, 2006. With permission.)

analytes after coating with the sinapinic acid solution by air-spray was minimal. The resulting MS imaging distribution of neither clozapine nor norclozapine was skewed by overlapping with other compound signals.

2.5 APPLICATIONS TO DRUG DISCOVERY AND DEVELOPMENT

MALDI–IMS is an excellent tool for visualizing small molecules in tissue sections. It is a more attractive option than other imaging modalities requiring labeled analytes, such as autoradiography and fluorescence imaging, because of its molecular specificity. The application of IMS to drug discovery is particularly appealing because it provides the opportunity to detect the localization of a drug and any metabolites that differ in mass from the parent drug in one experiment. It has been used to image drug distribution in various animal tissues as well as whole body slices.

For small molecules, the CAD fragmentation of the protonated molecules was usually performed to reduce or eliminate interference signals from the tissue or matrix [21, 65–68]. For example, one study that imaged a drug in tissue used a drug with a calculated protonated monoisotopic molecular weight that differed from the molecular weight of a sinapinic acid matrix cluster ion by less than 0.2 amu [21]. In this case, CAD was employed to fragment the drug ion at *m/z* 695 into a dominant fragment ion at *m/z* 228. The sinapinic acid cluster ion at *m/z* 695 fragmented into noninterfering ions at *m/z* 246 and *m/z* 471. Thus, MS/MS was used to localize the parent compound in a mouse tumor sample from a mouse dosed at 80 mg/kg.

MALDI–MS imaging was reliable and has been verified with autoradiography [21]. Another example using MALDI–MS/MS to image drugs in tissue was a study on clozapine levels in rat brain tissue in which results were compared with autoradiography studies. An autoradiograph taken from a rat brain following dosing with ^{3}H-labeled clozapine revealed that the drug was distributed throughout the rat brain, with the highest concentration in the lateral ventricle (Figure 2.10). A MALDI–MS/MS image of the clozapine fragment ion (*m/z* 327 → *m/z* 192) strongly correlated with the autoradiograph, showing the most intense signal in the ventricle area.

The MALDI instrument utilizes a relatively large target plate sufficient to contain a mouse whole-body tissue slice. Recently, some research teams have extended the applications of MALDI–IMS from an isolated organ tissue to whole-body sections [59]. Distinguishing drugs from metabolites has been demonstrated using MALDI–MS/MS imaging in whole-body rat sections. Olanzapine, an antipsychotic drug, was administered at a physiologically relevant dose to ten-week-old rats (given orally at 8 mg/kg). The animals were euthanized at 2 and 6 h post-dose. Whole-body sections were obtained on acetate film tape and mounted onto MALDI plates using double-sided conductive tape. MALDI–MS/MS analyses were performed on the whole-body sections to detect simultaneously olanzapine (*m/z* 313 → *m/z* 256) and two metabolites, *N*-desmethyl-olanzapine (*m/z* 299 → *m/z* 256) and 2-hydroxymethyl-olanzapine (*m/z* 329 → *m/z* 272). As shown in Figure 2.11, all three compounds can be detected in almost all tissues at both 2 and 6 h post-dose. Olanzapine was detected throughout the body at 2 h, and was present in the target organs, the brain and spinal cord. However, by 6 h it was significantly decreased in intensity in the brain and spinal cord (66% less ion signal). As expected from the results of classical studies, neither metabolite was detected in the brain or spinal cord. Both metabolites were present to a high degree in the liver and bladder. These data were also consistent with previous autoradiographic data and with the known metabolic pathways of rats [59]. This application highlights the ability of the technology to provide unambiguous localization information for several small molecules simultaneously.

There are also several successful examples of using MALDI–MS imaging for small-molecule analysis of the protonated molecule without CAD fragmentation [19, 69]. One study used chlorisondamine, a neuronal nicotinic ganglionic blocker, and cocaine administered to rats [69]. Both compounds were detected in the brains of rats via MALDI–IMS using α-cyano-4-hydroxycinnamic acid (CHCA) and 2,6- dihydroxybenzoic acid (DHB) as the matrices for chlorisondamine and cocaine, respectively. However, MS/MS was employed to confirm the identity of the protonated ions. Another study was reported by Bunch et al. to map the distribution of a xenobiotic substance, ketoconazole, in skin [19]. In this case, the detection of the protonated drug compound (along with diagnostic ^{37}Cl isotope signals) enabled the drug to be localized on the tissue surface and also localized by depth following the analysis of a cross-section.

These studies illustrate that MALDI–MS imaging can be performed in several ways to detect small molecules. Care must be taken when using MS to ensure there are no interfering peaks from the matrix or from endogenous compounds present in the tissue.

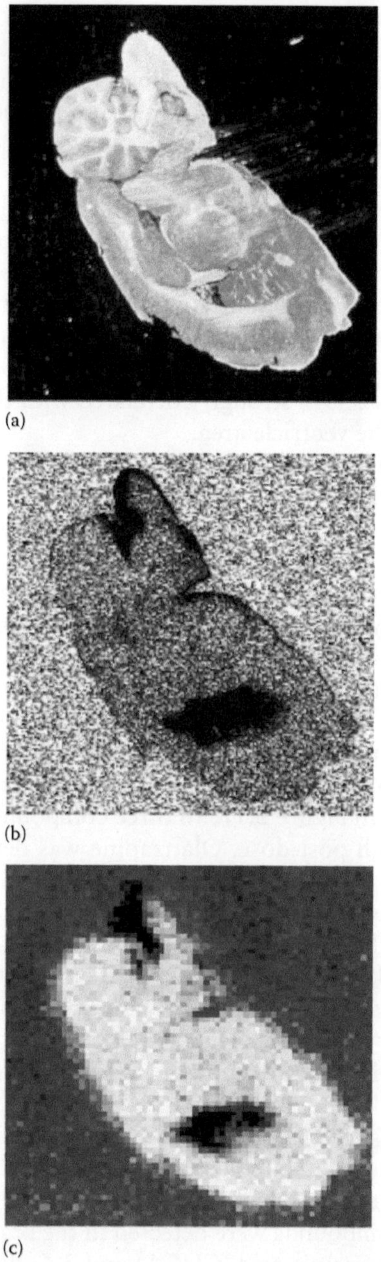

FIGURE 2.10 (a) Optical images, (b) radioautographic images, and (c) MALDI–MS/MS images from the study of rat brain tissue section. (From Hsieh, Y. et al., *Rapid Commun. Mass Spectrom.*, 20, 965, 2006. With permission.)

In addition, the specific fragmentation information obtained from MS/MS experiments adds a high degree of confidence in the identification of a signal as coming from the analyte of interest. However, the MS-only mode can give a multiplex advantage, in that many signals are detected at the same time from the same pixel. Sample preparation procedures, including the choice of matrix, extraction solvent, and detection method, must

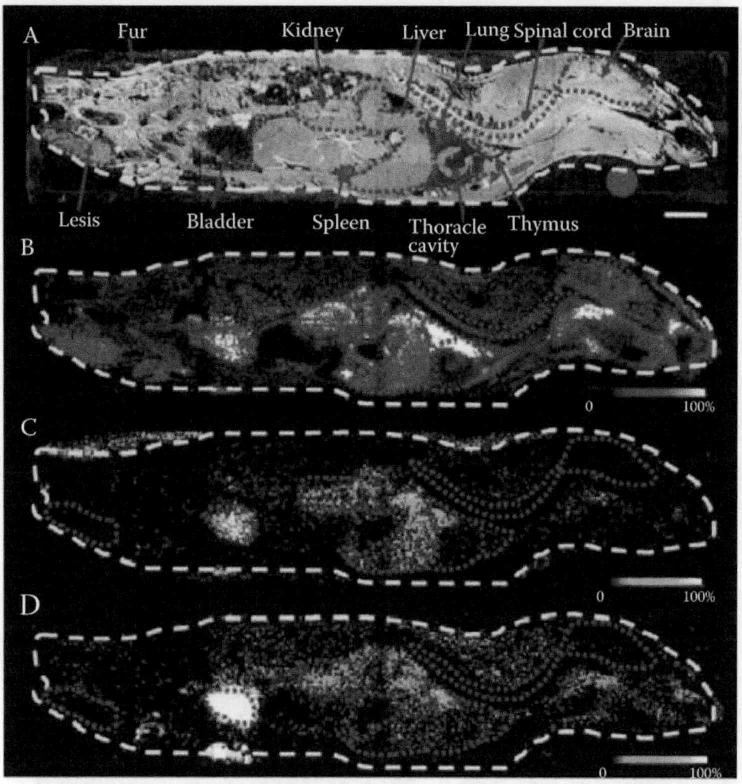

FIGURE 2.11 Detection of drug and metabolite distribution at 2 h post-dose in a whole rat sagittal tissue section by a single IMS analysis. Optical image of a 2 h post-OLZ dosed rat tissue section across four gold MALDI target plates. (A) Optical image (B) Organs outlined in red. Pink dot used as time point label. MS/MS ion image of OLZ (*m/z* 256). (C) MS/MS ion image of *N*-desmethyl metabolite (*m/z* 256). (D) MS/MS ion image of 2-hydroxymethyl metabolite (*m/z* 272). Bar, 1 cm. (From Khatib-Shahidi, S. et al., *Anal. Chem.*, 78, 6448, 2006. With permission.)

be optimized for each drug because of the wide variety of structures, solubilities, and physicochemical properties of drug compounds.

2.6 APPLICATION TO IMAGING OF PHOSPHOLIPIDS

Phospholipids play a very important role in many essential biological functions, including cell signaling, energy storage, and membrane structure and function. The analysis of lipids is challenging because of the wide molecular diversity of this class of compounds and their relative insolubility in aqueous systems. Using multiple stages of mass spectrometry [13, 70, 71], ion species may be unambiguously identified depending on how the lipid ion initially fragments. The use of ion mobility (IM) has also been reported in conjunction with MALDI–MS as a means to fractionate lipids [72–74] (Figure 2.12). In this technique, ions are first separated by IM and then analyzed by TOF mass spectrometry. The desorbed lipid ions fall on a trend line that is separate from those of oligonucleosides, peptides, proteins, and drugs and metabolites having the same nominal mass. This allows

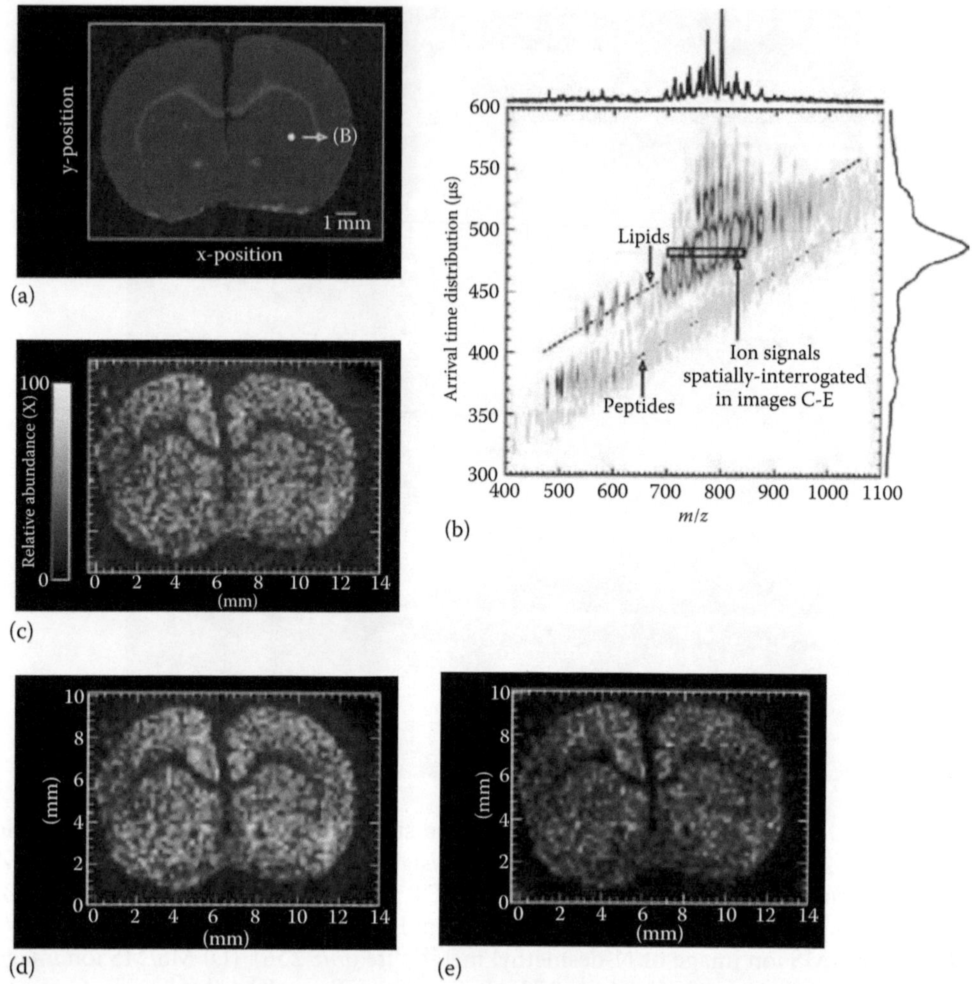

FIGURE 2.12 Imaging IM–MS for mapping the spatial coordinates of analytes based on structure and *m/z*. (a) An optical image of a thin coronal tissue section (12 μm) of a rat brain adjacent to the section used for imaging IM–MS. The spatial coordinates of the white spot indicated in (a) were interrogated to yield an IM–MS molecular profile as indicated in (b). Lipid imaging of the analytes indicated by the rectangle about an IM arrival time distribution of 480–484 μs and 700–840 *m/z* is illustrated in (c). (d, e) Extracted ion density maps for two phospholipids at *m/z* 771–776 [PC 32:0 + K]$^+$ and 819–823 [PC 32:0 + K]$^+$, in (d) and (e), respectively. MALDI was performed at a laser repetition rate of 1000 Hz with a spatial resolution of 200 μm. *Note* that the images in (c), (d), and (e) represent raw data which have not been smoothed or interpolated. The relative abundance scale in (c) is the same for (d) and (e). (From McLean, J.A. et al., *J. Mass Spectrom.*, 42, 1099, 2007. With permission.)

ions originating from lipids to be distinguished from other small molecules. In contrast with IMS, imaging IM–MS provides several unique advantages [72–74] including (1) the selective imaging of isobaric species or structural/conformational subpopulations of the same species on the basis of IM; (2) the separation/rejection of undesirable endogenous chemical noise; (3) the reduction of ion suppression effects in the source of the TOFMS, by temporal IM separation of analytes; and (4) the potential utility for nearly simultaneous IM–MS/MS of all analytes at a particular pixel coordinate.

2.7 CONCLUSIONS

During the past several years, significant progress has been made for the MALDI–MS imaging technique. It has been proven to be a valuable tool for studying drug distribution in animal tissues. MALDI–IMS is a simple, quick, and molecularly specific analytical platform that can distinguish small molecules from their metabolites on tissue sections. This technique is still in a very early stage of development. There is a need for further improvement in instrumentation and sample preparation. The use of new ion sources [75], new matrices [76], improved lasers [77], and advanced imaging software will improve our current ability to profile tissues by expanding the technological capabilities to decrease analysis time, and to increase sensitivity and versatility. The collaboration between analytical scientists and biological scientists will result in better experiment design and result interpretation. The potential of MALDI–IMS is great, and advances in the instrumentation and operating protocols will bring new applications and insights into molecular processes involving health and disease.

REFERENCES

1. Hsieh, Y. et al., High performance liquid chromatography–atmospheric pressure photoionization/tandem mass spectrometric analysis for small molecules in plasma, *Anal. Chem.*, 75, 3122, 2003.
2. Hsieh, Y. et al., Direct plasma analysis of drug compounds using monolithic column liquid chromatography and tandem mass spectrometry, *Anal. Chem.*, 75, 1812, 2003.
3. Korfmacher, W.A., Lead optimization strategies as part of a drug metabolism environment, *Curr. Opin. Drug Discov. Develop.*, 6, 481, 2003.
4. Korfmacher, W.A., Bioanalytical analysis in a drug discovery environment. In W.A. Korfmacher (Ed.), *Using Mass Spectrometry for Drug Metabolism Studies*, Boca Raton, FL: CRC Press, pp. 1–34, 2005.
5. Korfmacher, W.A., Principles and applications of LC-MS in new drug discovery, *Drug Discov. Today*, 10, 1357, 2005.
6. Chaurand, P. et al., Imaging mass spectrometry: Principles and potentials, *Toxicol. Pathol.*, 33, 92, 2005.
7. Rohner, T.C., Staab, D., and Stoeckli, M., MALDI mass spectrometric imaging of biological tissue sections, *Mech. Ageing Dev.*, 126, 177, 2005.
8. Rubakhin, S.S. et al., Imaging mass spectrometry: Fundamentals and applications to drug discovery, *Drug Discov. Today*, 10, 823, 2005.
9. Rudin, M., Rausch, M., and Stoeckli, M., Molecular imaging in drug discovery and development: Potential and limitations of nonnuclear methods, *Mol. Imaging Biol.*, 7, 5, 2005.
10. Chaurand, P., Schwartz, J., and Caprioli, R.M., Profiling and imaging proteins in tissue sections by MS, *Anal. Chem.*, 76, 87, 2004.
11. Stoeckli, M. et al., Imaging mass spectrometry: A new technology for the analysis of protein expression in mammalian tissues, *Nat. Med.*, 7(4), 493, 2001.

12. Todd, P.J. et al., Organic ion imaging of biological tissue with secondary ion mass spectrometry and matrix-assisted laser desorption/ionization, *J. Mass Spectrom.*, 36, 355, 2001.
13. Cornett, D.S. et al., MALDI imaging mass spectrometry: Molecular snapshots of biochemical systems, *Nat. Methods*, 4(10), 828, 2007.
14. Kang, M., Tholey, A., and Heinzle, E., Quantitation of low molecular mass substrates and products of enzyme catalyzed reactions using matrix-assisted laser desorption/ionization time-of-flight mass spectrometry, *Rapid Commun. Mass Spectrom.*, 14, 1972, 2000.
15. Duncan, M.W., Matanovic, G., and Cerpa-Poljak, A., Quantitative analysis of low molecular weight compounds of biological interest by matrix-assisted laser desorption ionization, *Rapid Commun. Mass Spectrom.*, 7, 1090, 1993.
16. Chen, Y. and Ling, Y., Detection of water-soluble vitamins by matrix-assisted laser desorption/ionization time-of-flight mass spectrometry using porphyrin matrices, *J. Mass Spectrom.*, 37, 716, 2002.
17. Donegan, M. et al., Sample preparation methods for MALDI analysis of small molecule metabolites, *Proceedings of the 51st ASMS Conference*, Montreal, QC, Canada, 2003.
18. Ayorinde, F.O. et al., Use of meso-tetrakis(pentafluorophenyl)porphyrin as a matrix for low molecular weight alkylphenol ethoxylates in laser desorption/ionization time-of-flight mass spectrometry, *Rapid Commun. Mass Spectrom.*, 13, 2474, 1999.
19. Bunch, J., Clench, M.R., and Richards, D.S., Determination of pharmaceutical compounds in skin by imaging matrix-assisted laser desorption/ionisation mass spectrometry, *Rapid Commun. Mass Spectrom.*, 18, 3051, 2004.
20. Crossman, L. et al., Investigation of the profiling depth in matrix-assisted laser desorption/ionization imaging mass spectrometry, *Rapid Commun. Mass Spectrom.*, 20, 284, 2006.
21. Hsieh, Y. et al., Matrix-assisted laser desorption/ionization imaging mass spectrometry for direct measurement of clozapine in rat brain tissue, *Rapid Commun. Mass Spectrom.*, 20, 965, 2006.
22. Reyzer, M.L. et al., Direct analysis of drug candidates in tissue by matrix-assisted laser desorption/ionization mass spectrometry, *J. Mass Spectrom.*, 38, 1081, 2003.
23. Troendle, F.J., Reddick, C.D., and Yost, R.A., Detection of pharmaceutical compounds in tissue by matrix-assisted laser desorption/ionization and laser desorption/chemical ionization tandem mass spectrometry with a quadrupole ion trap, *J. Am. Soc. Mass Spectrom.*, 10, 1315, 1999.
24. Hsieh, Y. et al., Mapping pharmaceuticals in tissues using MALDI imaging mass spectrometry, *J. Pharmacol. Toxicol. Methods*, 55, 193, 2006.
25. Boy, C. et al., Imaging dopamine D_4 receptors in the living primates brain: A positron emission tomography study using the novel D_1/D_4 antagonist [^{11}C]SDZ GLC 756, *Synapse*, 30, 341, 1998.
26. Ekesbo, A. et al., Effects of the substituted (S)-3-phenylpiperidine (-)-OSU6162 on PET measurements of [11C]SCH23390 and [11C]raclopride binding in primate brains, *Neuropharmacology*, 38, 331, 1999.
27. Saleem, A. et al., Modulation of fluorouracil tissue pharmacokinetics by eniluracil: In vivo imaging of drug action, *Lancet*, 355, 2125, 2000.
28. Moresco, R.M. et al., PET in psychopharmacology, *Pharmacol. Res.*, 44(3), 151, 2001.

29. Ishiwata, K. et al., Positron emission tomography and ex vivo and in vitro autoradiography studies on dopamine D_2-like receptor degeneration in the quinolinic acid-lesioned rat striatum: Comparison of [^{11}C]raclopride, [^{11}C]nemonapride, and [^{11}C]*n*-methylspiperone, *Nucl. Med. Biol.*, 29, 307, 2002.
30. Kassiou, M. et al., (+)-[^{76}Br]A-69024: A non-benzazepine radioligand for studies of dopamine D_1 receptors using PET, *Nucl. Med. Biol.*, 29, 295, 2002.
31. Aboagye, E.O. and Price, P.M., Use of positron emission tomography in anticancer drug development, *Invest. New Drugs*, 21, 169, 2003.
32. Singh, M. and Walnuch, V., Physics and instrumentation for imaging in vivo drug distribution, *Adv. Drug Deliv. Rev.*, 41, 7, 2000.
33. Port, R.E. and Wolf, W., Noninvasive methods to study drug distribution, *Invest. New Drugs*, 21, 157, 2003.
34. Faas, H. et al., Monitoring the intragastric distribution of a colloidal drug carrier model by magnetic resonance imaging, *Pharm. Res.*, 18(4), 460, 2001.
35. Kupriyanov, V.V. et al., The effects of drugs modulating K+ transport on Rb+ uptake and distribution in pig hearts following regional ischemia: 87Rb MRI study, *NMR Biomed.*, 15, 348, 2002.
36. Noworolski, S.M. et al., High spatial resolution ^{1}H-MRSI and segmented MRI of cortical gray matter and subcortical white matter in three regions of the human brain, *Magnet. Reson. Med.*, 41, 21, 1999.
37. Solon, E.G. and Kraus, L., Quantitative whole-body autoradiography in the pharmaceutical industry: Survey results on study design, methods, and regulatory compliance, *J. Pharmacol. Toxicol.*, 46, 73, 2002.
38. Coe, R.A.J., Quantitative whole-body autoradiography, *Regul. Toxicol. Pharm.*, 31, S1, 2000.
39. Langlois, X. et al., Use of the beta-imager for rapid ex vivo autoradiography exemplified with central nervous system penetrating neurokinin 3 antagonists, *J. Pharmacol. Exp. Ther.*, 299(2), 712, 2001.
40. Tribollet, E. et al., Localization and quantitation of tritiated compounds in tissue sections with a gaseous detector of beta particles: Comparison with film autoradiography, *PNAS*, 88, 1466, 1991.
41. Lees, J.E. et al., An MCP-based system for beta autoradiography, *IEEE Trans. Nucl. Sci.*, 46(3), 636, 1999.
42. Lees, J.E., Fraser, G.W., and Carthew, P., Microchannel plate detectors for ^{14}C autoradiography, *IEEE Trans. Nucl. Sci.*, 45(3), 1288, 1998.
43. McDonnell, L.A. and Heeren, R.M.A., Imaging mass spectrometry, *Mass Spectrom. Rev.*, 26, 606, 2007.
44. Pacholski, M.L. and Winograd, N., Imaging with mass spectrometry, *Chem. Rev.*, 99, 2977, 1999.
45. Pacholski, M.L. et al., Static time-of-flight secondary ion mass spectrometry imaging of freeze-fractured, frozen-hydrated biological membranes, *Rapid Commun. Mass Spectrom.*, 12, 1232, 1998.
46. Audinot, J.-N. et al., Detection and quantification of benzodiazepines in hair by TOF-SIMS: Preliminary results, *Appl. Surf. Sci.*, 203–204, 718, 2003.
47. Winograd, N., Prospects for imaging TOF-SIMS: From fundamentals to biotechnology, *Appl. Surf. Sci.*, 203–204, 13, 2003.
48. Guerquin-kern, J.-L. et al., Complementary advantages of fluorescence and SIMS microscopies in the study of cellular localization of two new antitumor drugs, *Microsc. Res. Techniq.*, 36, 287, 1997.

49. Chandra, S., Lorey II, D.R., and Smith, D.R., Quantitative subcellular secondary ion mass spectrometry (SIMS) imaging of boron-10 and boron-11 isotopes in the same cell delivered by two combined BNCT drugs: In vitro studies on human glib-lastoma T98G cells, *Radiat. Res.*, 157, 700, 2002.
50. Fragu, P. and Kahn, E., Secondary ion mass spectrometry (SIMS) microscopy: A new tool for pharmacological studies in humans, *Microsc. Res. Techniq.*, 36, 296, 1997.
51. Todd, P.J. et al., Organic ion imaging of biological tissue with secondary ion mass spectrometry and matrix-assisted laser desorption/ionization, *J. Mass Spectrom.*, 36, 355, 2001.
52. McCandlish, C.A., McMahon, J.M., and Todd, P.J., Secondary ion images of the rodent brain, *J. Am. Soc. Mass Spectrom.*, 11, 191, 2000.
53. Todd, P.J. et al., Organic SIMS of biologic tissue, *Anal. Chem.*, 69, 529A, 1997.
54. Wong, S.C.C. et al., Development of a $C60^{+}$ ion gun for static SIMS and chemical imaging, *Appl. Surf. Sci.*, 203–204, 219, 2003.
55. Caprioli, R.W., Farmer, T.B., and Gile, J., Molecular imaging of biological samples: Localization of peptides and proteins using MALDI-TOF MS, *Anal. Chem.*, 69, 4751, 1997.
56. Hsieh, Y., Chen, J., and Korfmacher, W.A., Mapping pharmaceuticals in tissues using MALDI imaging mass spectrometry, *J. Pharmacol. Toxicol. Methods*, 55(2), 193, 2007.
57. Schwartz, S.A., Reyzer, M.L., and Caprioli, R.M., Direct tissue analysis using matrix assisted laser desorption/ionization mass spectrometry: Practical aspects of sample preparation, *J. Mass Spectrom.*, 38, 699, 2003.
58. Chaurand, P. et al., New developments in profiling and imaging of proteins from tissue sections by MALDI mass spectrometry, *J. Proteome Res.*, 5, 2889, 2006.
59. Khatib-Shahidi, S. et al., Direct molecular analysis of whole-body animal tissue sections by imaging MALDI mass spectrometry, *Anal. Chem.*, 78, 6448, 2006.
60. Shimma, S. et al., Direct MS/MS analysis in mammalian tissue sections using MALDI-QIT-TOFMS and chemical inkjet technology, *Surf. Interface Anal.*, 38, 1712, 2006.
61. Aerni, H.R. et al., Automated acoustic matrix deposition for MALDI sample preparation, *Anal. Chem.*, 78, 827, 2006.
62. Ohkawa, T. and Bunch, J., MALDI MS imaging to reveal distribution of benzodiazepine drug and metabolite molecules in rat brain, *Proceedings of the 55th ASMS Conference*, Indianapolis, IN, 2007.
63. Cohen, S.L. and Chait, B.T., Influence of matrix solution conditions on the MALDI-MS analysis of peptides and proteins, *Anal. Chem.*, 68, 31, 1996.
64. Dreisewerd, K.L. et al., Influence of the laser intensity and spot size on the desorption of molecules and ions in matrix-assisted laser desorption/ionization with a uniform beam profile, *Int. J. Mass Spectrom.*, 141, 127, 1995.
65. Stoeckli, M. et al., Compound and metabolite distribution measured by MALDI mass spectrometric imaging in whole-body tissue sections, *Int. J. Mass Spectrom.*, 260, 195, 2007.
66. Reyzer, M.L. and Caprioli, R.M., MALDI-MS-based imaging of small molecules and proteins in tissues, *Curr. Opin. Chem. Biol.*, 11, 29, 2007.
67. Pevsner, P.H. et al., Direct identification of proteins from T47D cells and murine brain tissue by matrix-assisted laser desorption/ionization postsource decay/collision-induced dissociation, *Rapid Commun. Mass Spectrom.*, 21, 429, 2007.

68. Atkinson, S.J. et al., Examination of the distribution of the bioreductive drug AQ4N and its active metabolite AQ4 in solid tumours by imaging matrix-assisted laser desorption ionization mass spectrometry, *Rapid Commun. Mass Spectrom.*, 21, 1271, 2007.
69. Wang, H.-Y.J. et al., Localization and analyses of small drug molecules in rat brain tissue sections, *Anal. Chem.*, 77, 6682, 2005.
70. Garrett, T.J. et al., Analysis of intact tissue by intermediate-pressure MALDI on a linear ion trap mass spectrometer, *Anal. Chem.*, 78, 2465, 2006.
71. Cha, S.W. et al., Colloidal graphite-assisted laser desorption/ionization mass spectrometry and MS^n of small molecules. 1. Imaging of cerebrosides directly from rat brain tissue, *Anal. Chem.*, 79, 2373, 2007.
72. Woods, A.S. et al., IR-MALDI-LDI combined with ion mobility orthogonal time-of-flight mass spectrometry, *J. Preoteome Res.*, 5, 1484, 2006.
73. Jackson, S.N. et al., Direct tissue analysis of phospholipids in rat brain using MALDI-TOFMS and MALDI-ion mobility-TOFM, *J. Am. Soc. Mass Spectrom.*, 16, 133, 2005.
74. McLean, J.A. et al., Profiling and imaging of tissues by imaging ion mobility-mass spectrometry, *J. Mass Spectrom.*, 42, 1099, 2007.
75. Wiseman, J.M. et al., Mass spectrometric profiling of intact biological tissue by using desorption electrospray ionization, *Angew Chem. Int. Ed. Engl.*, 44, 7094, 2005.
76. Lemaire, R. et al., Solid ionic matrixes for direct tissue analysis and MALDI imaging, *Anal. Chem.*, 78, 809, 2006.
77. Holle, A. et al., Optimizing UV laser focus profiles for improved MALDI performance, *J. Mass Spectrom.*, 41, 705, 2006.

SECTION 2

Different Methodologies

CHAPTER 3

Matrix-Assisted Laser Desorption Ionization (MALDI) and Atmospheric Pressure-MALDI (AP-MALDI)

Nrusingha Charan Basantia

CONTENTS

3.1 INTRODUCTION

Matrix-assisted laser desorption/ionization (MALDI) mass spectrometry is one of the most widely used soft ionization methods for biomolecules (Tanaka et al., 1988; Karas and Hillenkamp, 1988). Though there exists a number of ionization methods used for mass spectrometry, such as chemical ionization (CI), electron ionization (EI), inductively coupled plasma (ICP), and desorption electrospray ionization (DESI), and many more, few of them are capable of soft ionization of macromolecules like proteins, carbohydrates, and lipids. MALDI is, giving it a distinct niche. Several unique aspects of MALDI, such as the production of primarily singly charged species and the high tolerance against contaminants, make MALDI a promising tool for exploring unknown structures in a complex system. It has become one of the most well-established approaches in proteomics owing to its high sensitivity, low sample consumption, and minimal sample handling. With the growing interest in biological roles of carbohydrates, the structural identification of carbohydrates has also been attempted by MALDI mass spectrometry. More recently, atmospheric pressure-MALDI (AP-MALDI) was introduced, increasing the ease of sample preparation and allowing for the analysis of volatile molecules, as the sample no longer needs to be placed under vacuum prior to analysis (Laiko et al., 2000a, b). The application of the MALDI technique alone or in combination with mass spectrometry imaging (MSI) is gaining importance in the analysis of secondary metabolites in plant species to identify the quality, variety, and authenticity of food grains, the geographical origin of fruits and vegetables, and the authenticity of meat and meat products.

3.2 THEORY AND MECHANISM

MALDI in brief comprises of photochemical (ionization and fragmentation) and mechanical (desorption) events. In MALDI, analytes are incorporated into organic matrices in a solid or viscous phase (ionic liquid matrices). A pulsed laser irradiates the sample to facilitate the ionization/desorption reactions. During laser excitation, matrices take up and redistribute the laser energy for the desorption and ionization of analytes.

3.2.1 Formation of Matrix Analyte Co-Crystal

The first step of MALDI requires the formation of a matrix analyte co-crystal. A sample of analyte is dissolved into a solution of the matrix compound and spotted onto a plate. The solvents then evaporate, producing a recrystallized matrix with interstitial analyte particles; this substance is known as a co-crystal.

3.2.1.1 Matrix

MALDI is based on the use of a matrix that absorbs the energy of the laser irradiation and is, thus, crucial for the successful generation of ions. The choice of matrix compound is the key to obtaining a quality spectrum. There are basically three different classes of matrices available (Leopold et al., 2018):

i. Classical organic matrices (e.g. benzoic acid or cinnamic acid derivatives)
ii. Liquid crystalline matrices
iii. Inorganic matrices (e.g. graphite that provides only a weak background)

An ideal MALDI matrices should fulfil the five following criteria (Horneffer et al., 1999):

i. The matrix must possess a strong absorption at the emission wavelength of the laser typically, in the UV range at either 337 or 355 nm.
ii. A useful matrix ensures the ion formation of the analyte. Since the often-used aromatic ring system is poorly soluble in polar solvent systems, carboxylic acid (normally benzoic acid or cinnamic acid derivatives) is an often-used structure of matrix compounds. Since the carboxylic group is both polar and acidic, it ensures the solubility of the matrix in polar solvents and the protonation of the analyte, respectively.

iii. The matrix should be stable under high vacuum conditions for as long as possible. Although this seems trivial, there are many promising MALDI matrices (e.g. 4-nitroaniline (PNA) or 2,6-dihydroxyacetophenone) which fail to fulfil this condition. The MALDI ionization process occurs under high vacuum conditions, and many compounds undergo a sublimation process. This leads to a continuous change of the matrix/analyte ratio, and this may be one reason why the MALDI mass spectra show time-dependent changes.
iv. A perfect matrix should isolate the generated ions and prevent the generation of analyte clusters, for instance, dimer formation. Such clusters would complicate the spectra and decrease the sensitivity. Cluster avoidance is the primary reason why a significant excess of the matrix over the analyte should be used.
v. The crystallization between the matrix and the analyte leads to co-crystals which should be as homogeneous as possible. The improvement of the crystal homogeneity is very important since this determines the "shot-to-shot" reproducibility of the acquired MALDI mass spectra. (O'Rourke et al., 2018 and Szájli et al., 2008).

In Table 3.1 the structures of frequently used MALDI matrices, their chemical names, trivial names, and abbreviations are given.

3.2.1.2 Laser Light

The majority of MALDI studies, so far, have been performed with ultraviolet (UV) lasers. We will focus nearly exclusively on this laser type, with an emission wavelength of either 337 nm or more recently 355 nm. The first wavelength is typically emitted by an N2 laser, while 355 nm is characteristic of a neodymium-doped yttrium aluminium garnet (Nd: YAG) laser. During the last few years, N2 lasers have been increasingly replaced by Nd: YAG since the N2 lasers have a shorter lifetime, compared to Nd: YAG lasers, and, moreover, the photon energy at 355 nm is lower than at 337 nm. This is beneficial if labile compounds have to be analysed.

Nevertheless, IR-MALDI MS is also quite useful and offers some important advantages. When the pulsed laser beam hits the sample (normally co-crystals between the matrix and the analyte), its energy is primarily absorbed by the matrix that is present in an, approximately, thousand-fold excess over the analyte and has a significant absorption coefficient at the laser wavelength. Consequently, the matrix is vaporized, carrying the intact analyte molecules into the vapour phase. It has been suggested that MALDI is such a soft process because (among other reasons) the sample is expanded from the solid state into a high vacuum, leading to a significant cooling of the sample.

3.2.2 Ion Production in MALDI

Different mechanisms have been proposed to explain MALDI ionization (Karas et al., 1985). All are in agreement with the presence of two different steps: (i) primary ion formation from the matrix (Ma); and (ii) secondary ion formation of the analyte (A) originating from the gas phase interaction of reactive matrix ions and neutral molecules of the analyte. The process of ion production is shown in Figure 3.1 (Aparna et al., 2015). However the photon energy for N_2 laser and Nd:Yag laser is much lower than the energy required for matrix ionization. Setz and Knochenmuss (2005) proposed that multiple photon absorption, leading to multiple electronic excited matrix molecules, followed by energy pooling phenomena, would lead to the formation of the matrix odd electron molecular ion Ma+• that, in turn, reacts with a neutral matrix molecule leading

TABLE 3.1 List of Frequently Used MALDI Matrices

Sl. No	Compound	Acronym	Application to	Reference
1	Nicotinic acid	NA	Peptides, proteins	Dougherty et al., 1978
2	Picolinic acid	PA	Oligonucleotides, DNA	
3	Nicotinic acid	NA	Peptides, proteins	
5	3-Hydroxypicolinic acid	HPA, 3-HPA	Oligonucleotides, DNA	
6	3-Aminopicolinic acid	3-APA	Oligonucleotides, DNA	
7	6-Aza-2-thiothymine	ATT	Oligonucleotides, DNA	
8	2,5-Dihydroxybenzoic acid	DHB	Proteins, oligosaccharides	
9	DHB-based mixtures	DHB/XY and super-DHB	Proteins, oligosaccharides	
10	3-Aminoquinoline	3-AQ	Oligosaccharides	
12	α-Cyano-4-hydroxycinnamic acid	α-CHC, α-CHCA, 4-HCCA, CHCA	Peptides, smaller proteins, triacyl-glycerols, numerous other compounds	
13	4-Chloro-α-cyano-cinnamic acid	ClCCA	Peptides	
14	3,5-Dimethoxy-4-hydroxycinnamic acid	SA	Proteins	
15	2-(4-Hydroxyphenylazo) benzoic acid	HABA	Peptides, proteins, glycoproteins, polystyrene	
16	2-Mercaptobenzothiazole	MBT	Peptides, proteins, synthetic polymers	
17	5-Chloro-2-mercaptobenzothiazole	CMBT	Glyco-peptides, phospho-peptides, and proteins	
18	2,6-Dihydroxyacetophenone	DHAP	Glyco-peptides, phosphor-peptides, proteins	
19	2,4,6-Trihydroxyacetophenone	THAP	Solid-supported oligonucleotides	
20	Dithranol (1,8,9-anthracenetriol)	None	Synthetic polymers	
21	9-Nitroanthracene	9-NA	Fullerenes and derivatives	
22	Benzo[a]pyrene	None	Fullerenes and derivatives	

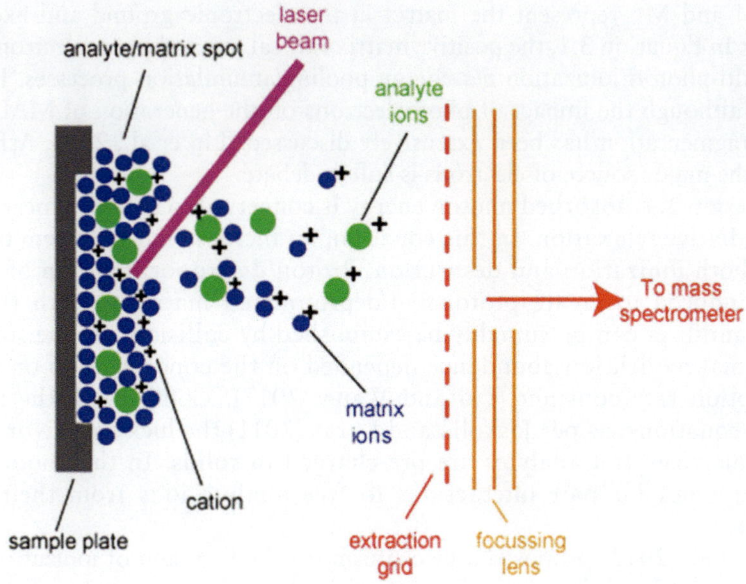

FIGURE 3.1 Process of ionization in MALDI-MS. (From Aparna et al., 2015.)

to [Ma + H]+. The last species must be considered responsible for the analyte (A) protonation in gas phase (Equations 3.1 and 3.2):

$$\mathbf{MaH + +A --\!\!\rightarrow Ma + AH+} \tag{3.1}$$

$$\text{only if } PA_A > PA_{Ma}, \ (PA = \text{Proton Affinity}). \tag{3.2}$$

The possible limitation of this mechanism lies in the low probability of multiphoton absorption and energy pooling, which makes the high ionic yield observed in MALDI experiments difficult to explain. Alternatively it has been considered that the electronic excitation (one photon required) leads in general to an increase of the matrix acidity, promoting proton transfer. However, excited-state proton transfer (ESPT) does not affect carboxylic acids, and consequently it cannot be invoked as effective mechanism for proton transfer for the MALDI matrices. Furthermore ESPT is active for compounds that do not exhibit any activity as MALDI matrix (Karas et al., 1985; Ireland and Wyatt, 1976). Other ionization mechanisms have been proposed such as laser-induced shockwaves (Dlott et al., 1998) or thermal ionization (Dreisewerd, 1995).

After photo-absorption, a two-step ionization process suggests that the ionization of matrices is an initial ionization step, followed by ion molecule reactions to produce charged analyte species (Hatakeyama and Tachikawa, 2011). However, how matrices are initially ionized remains open to debate. Two major arguments of postulated initial ionizations include firstly photoionization in the gas phase as delineated in Equation 3.3, and secondly thermal energy promoting ionizations in the condensed phase (Equation 3.4).

$$M + nh\nu \rightarrow \ M^* \rightarrow M^{+\cdot} \ + e^- \tag{3.3}$$

$$2M(s) + nh\nu \leftrightarrow 2M\ (^*s) \ \leftrightarrow [M+H]^+_{(s)} + [M-H]^-_{(s)} \tag{3.4}$$

In which M and M* represent the matrix in the electronic ground and excited states, respectively. In Equation 3.1, the positive matrix radical ion and photoelectrons are generated by multi-photon ionization *via* energy pooling/annihilation processes. It should be noted that, although the impact of photoelectrons on the generation of MALDI ion and trigger of fragmentation has been extensively discussed (Liu et al., 2010; Asfandiarov et al., 2002), the major source of electrons is still in debate.

In Equation 3.4, absorbed photon energy is converted to thermal energy *via* ultrafast non-radiative relaxation. In this equation, an increase in the system temperature facilitates both ionization and desorption. Proton disproportionation of matrices is thermally induced to create protonated/deprotonated matrices. Both thermal and chemical equilibria can presumably be established by collisions on the solid surface. In the thermal model, ion abundance depended on the concentration on the surface and desorption rate constant (Lai and Wang, 2017). Contrary to the two above-mentioned equations, as per Jaskolla and Karas (2011) the lucky survivor model suggests that matrices and analytes are pre-charged in solids. In this model, the laser energy overcomes ion-pair interactions to free analyte ions from their respective counter ions.

Mollin et al. (2012) proposed a mechanism for the first step of ionization occurring in matrix-assisted laser desorption ionization, leading to protonated and deprotonated matrix (Ma) molecules ([Ma + H]+ and [Ma – H]– ions). It is based on the observation that in solid state, for carboxyl-containing MALDI matrices, the molecules form strong hydrogen bonds and their carboxylic groups can act as both donors and acceptors. This behaviour leads to stable dimeric structures. The laser irradiation leads to the cleavage of these hydrogen bonds, and theoretical calculations show that both [Ma + H]+ and [Ma – H]– ions can be formed through a two-photon absorption process. It should be emphasized that a single ionization mechanism that is applied universally to elucidate all MALDI observations is highly unlikely owing to different photochemical properties of matrices as well as excitation conditions.

3.2.3 Desorption (Phase Transition)

Several theories have been developed to explain the desorption of large molecules by MALDI. The thermal-spike model proposes that the matrix molecules sublime from the surface as a result of local heating at low laser influence. Desorption in MALDI is the phase transition of samples from the condensed phase to the gas phase. It is a prerequisite for molecules to be detected by MALDI mass spectrometry. Desorption processes, including thermal desorption occurring at low and moderate laser energy density and phase explosion happening at high laser energy density, were studied quantitatively on both experimental and theoretical bases. Results of these studies suggest that factors affecting the desorption efficiency can be classified into two categories: excitation parameters (Hornefer et al., 1999) and intrinsic properties of samples (Jaskolla et al., 2009; Mowry and Jhonston, 1994; Schulz et al., 2006). Generally, desorption efficiency increases when MALDI crystal size decreases and the system temperature increases (Schulz et al., 2006). On the basis of factors influencing desorption efficiency, mathematical equations as well as computational methods, such as molecular dynamics (MD) simulations, have been used to quantitatively describe the desorption process (Zhigilei et al., 2003).

3.2.4 Mass Analyser – Time of Flight (TOF)

There are three types of mass analysers typically used with the MALDI ionization source: a linear time-of-flight (TOF), a TOF reflectron, and a Fourier transform mass analyser. The linear TOF mass analyser is the simplest of the three devices and has enjoyed a renaissance with the invention of MALDI. TOF analysis is based on accelerating a set of ions to a detector where all of the ions are given the same amount of energy. Because the ions have the same energy, but a different mass, the ions reach the detector at different times. The smaller ions reach the detector first because of their greater velocity while the larger ions take longer owing to their larger mass. Hence, the analyser is called TOF because the mass is determined from the ions' time of flight. The arrival time at the detector is dependent upon the mass, charge, and kinetic energy (KE) of the ion. Since KE is equal to 1/2 mv^2 or velocity (v) = $(2KE/m)^{1/2}$. Ions will travel a given distance, *d*, within a time, *t*, where *t* is dependent upon their mass-to-charge ratio (*m/z*).

The ions then accelerate in a TOF analyser because they are subject to an equal electric field. TOF is a field-free flight tube. The ions travel in a straight and linear direction to the detector. The mass to charge ratio (*m/z*) of the sample ions can be calculated using the Equation 3.5.

$$T = C1\ (m/z)0.5 + C2. \tag{3.5}$$

C1 and C2 are instrumental constants, which can be determined with compounds of known mass.

The TOF reflectron combines TOF technology with an electrostatic analyser, the reflectron. The reflectron serves to increase the amount of time, *t*, and ions need to reach the detector while reducing their KE distribution, thereby reducing the temporal distribution 1/*t*. Since resolution is defined by the mass of a peak divided by the width of a peak or *m*/1*m* (or *t* = 1*t* since *m* is related to *t*), increasing *t* and decreasing 1*t* results in higher resolution. This increased resolution, however, often comes at the expense of sensitivity and a relatively low mass range, typically <10,000 *m/z*. One innovation that has had a dramatic effect on increasing the resolving power of MALDI-TOF instruments is delayed extraction (DE). DE is a relatively simple means of cooling the ions (and possibly focusing them) immediately after the MALDI ionization event. In traditional MALDI instruments the ions are accelerated out of the ionization source as soon as they are formed. However, with delayed extraction the ions are allowed to "cool" for 150 nano-seconds before being accelerated to the analyser. This cooling period generates a set of ions with a much lower KE distribution, ultimately reducing the temporal spread of ions once they enter the TOF analyser. Overall, this results in increased resolution and accuracy.

3.2.5 Fragmentation

Fragmentation techniques are essential methods to deposit energy into a targeted ion to cause reproducible bond cleavages. Cleavages yield diagnostic and interpretable fragment ions that reveal structural or sequence information about the molecule of interest. Commonly used fragmentation techniques in MALDI mass spectrometry include laser-induced promptly in-source decay (ISD), post-source decay (PSD, a unimolecular dissociation of metastable ions), and collision-induced dissociation (CID). ISD ions are promptly produced by thermal activation and/or chemical reactions during desorption/ionization processes in the source region, such as hydrogen radical-mediated and/or free electron-mediated reactions. More energy input and higher energy-coupling efficiency between matrices and analytes enhance the ISD process. Because ISD ions receive the full acceleration energy in the ion source, they can be well-resolved in the mass spectra. On the other hand, PSD ions are typically generated on a time scale of micro- to milliseconds in the field-free region. Owing to having the same velocity as their precursor ions, PSD ions cannot be resolved in linear TOF instruments. The energy liberated by gas-phase proton transfer reactions can greatly influence fragmentation. For oligonucleotides, Zhu et al. (1995) have proposed that the loss of protonated base, followed by backbone fragmentation and disproportionation, ultimately leads to anionic fragments. Consequently, the difference in proton affinity between the analyte and matrix should determine the energy available. For reducing the fragmentation of oligo-deoxynucleotides, matrices with relatively high proton affinities should, therefore, be chosen (Zenobi and Knochenmuss, 1998).

3.3 AP MALDI

This apparatus provides an ion source using matrix-assisted laser desorption and ionization at or near atmospheric pressure. This versatile and simple apparatus has a non-destructive ion source.

To analyse samples with atmospheric-pressure ions any mass spectrometer must be equipped with an atmospheric pressure interface (API) to transfer ions from an external region of atmospheric pressure to a mass analyser under high vacuum. Ions produced under atmospheric pressure conditions may be used for other analytical purposes, too. For example, they are used in ion mobility spectroscopy (IMS), which is a fast-growing branch of analytical chemistry. An API is used to transfer ions from an atmospheric-pressure ion source, such as an ESI, to a vacuum of a mass spectrometer. This interface has an efficiency as low as a few percent. AP-MALDI has not been applied because of the concern that MALDI does not generate enough ions to compensate the loss of ions due to the API. However AP-MALDI has the characteristics of easy sample preparation, high stability, high contamination tolerance, interface with other analytical separation techniques, etc.

The AP-MALDI apparatus mainly consists of three parts:

i. An atmospheric pressure ionization chamber which hosts a sample to be analysed
ii. A laser system outside the ionization chamber for illuminating the sample in the ionization chamber
iii. An interface which connects the ionization chamber to the spectrometer

The ionization chamber is used to control the gas nature, pressure, temperature, and humidity. In some cases, additional equipment is incorporated in the ionization chamber to control these parameters, such as a heater to control the temperature.

Ionization chamber: In cases when the ionization process is conducted in ambient air, even the use of the ionization chamber is optional. The ionization chamber typically comprises a bath gas inlet as a pathway for the bath gas to enter the chamber. Normally, the ionization chamber is filled with a bath gas at or near atmospheric pressure. The bath gas, which is normally selected from the group which comprises inert gas, nitrogen gas, or a gas mix such as air, is chosen such that it does not react with the sample or by itself, even under laser illumination.

Laser system for illumination: The ionization chamber further comprises a window through which the illuminating laser beam enters. The position of the window is correlated to the position of the sample to be illuminated inside the ionization chamber. In a preferred embodiment, the window is positioned at the side of the chamber. The sample, also referred to as the target material, normally comprises a mixture of analyte materials and light-absorbing matrix substances. The sample is in a form selected from the group of solid phase and liquid phase. The sample is deposited on a target surface of a sample support. When illuminated with the laser beam, the matrix molecules are ionized and evaporated.

The ionized matrix molecules subsequently ionize the analyte molecules through the charge transfer process. At the same time, the analyte molecules, analyte ions, and fragmented analyte ions are evaporated together with the matrix ions and molecules. Examples of matrix substances are C-cyano-4-hydroxycinnamic acid, sinapinic acid, and 3-hydroxypicolinic acid.

3.3.1 Difference between MALDI and AP-MALDI

AP-MALDI takes place under atmospheric pressure conditions. This allows a more or less uniform ion cloud to form after laser illumination, because the produced ions achieve a thermal equilibrium with the surrounding bath gas molecules quickly through collision. As a consequence, the AP-MALDI technique produces a quasi-continuous ion source which provides a stable ion supply to the spectrometer. A more powerful laser pulse is used in AP-MALDI because vibrationally excited analyte ions are quickly thermalized (stabilized) with the surrounding bath gas molecules before they dissociate into fragments. Furthermore, a larger laser spot is used to illuminate the sample, which allows an easier alignment procedure in comparison with the vacuum MALDI technique. As a consequence, a substantial number of ions, as much as a few picomoles, are generated in AP-MALDI to compensate for the loss due to API. AP-MALDI has an ion source which is external with respect to the spectrometer instrument. Thus any mass spectrometer equipped with an atmospheric pressure interface (API) may be easily coupled with this ion source without undue effort. The de-coupling of the ion source from the ion focusing optics of a spectrometer ensures the same resolution level and spectra calibration procedure as for any other atmospheric pressure ionization technique. As a result, other atmospheric pressure separation techniques, such as ion mobility spectroscopy, may be easily coupled with AP-MALDI. The atmospheric pressure character of AP-MALDI allows a simple sample loading procedure. Consequently, the construction of the instrument is simplified drastically. Both sample preparation and ionization processes take place under atmospheric pressure conditions. This enables a simple and straightforward way for the on-line coupling of AP-MALDI with separation techniques such as HPLC and CZE. AP-MALDI is a versatile technique. The selection of a possible matrix material for AP-MALDI is not limited to solid or liquid matrices with very low vapour pressures. Matrices of volatile liquids may be used under atmospheric pressure conditions. Furthermore, AP-MALDI achieves the ionization and desorption of the analyte in a single step. This property of AP-MALDI allows simple equipment construction and operation, which also makes AP-MALDI advantageous over prior art. Prior art relies on a twostep process: a laser beam decomposes matrix molecules in order to release the analytes; the released analyte is subsequently ionized by an atmospheric pressure chemical ionization process.

3.4 APPLICATIONS

3.4.1 MALDI and Rice, Grains, and Seed

MALDI MSI as a versatile imaging technique has been widely introduced in plant sciences mostly focusing on the endogenous molecules, including primary and secondary metabolites as well as peptides and proteins. The correlation of molecular information to anatomical and histochemical images in plant tissues offers tremendous potential for a better understanding of various plant physiological processes. As rice has been a major agricultural product in Asia the commercial value of specific varieties of rice grown in good conditions can be very expensive. This difference in commercial value leads to the incidence of falsely labelling ordinary rice as high-grade rice. Therefore there is a dire need for a simple and reliable method to distinguish between ordinary and high-grade rice. MALDI-TOF-MS is a suitable, quick, and simple method for the evaluation of rice varieties based on the fingerprints of phytochemicals.

As the seed is a very important part of the plant it is desirable to understand the relationship between endogenous molecules in seeds and how metabolic processes are controlled in the seed cells. It is necessary to determine the amount and distribution of endogenous molecules present in the seed cell. The task of identifying and mapping plant endogenous molecules is still a huge challenge. MALDI-MSI as a novel technique shows great potential in solving the problem (Bjarnholt et al., 2014). Various MALDI-MSI studies have been performed for endogenous molecules in various seeds such as wheat, rice, cotton seeds, and corn, and for different molecules such as amino acids, sugars, phosphorylated metabolites, nucleotides, and glycosides (Table 3.2).

3.4.2 Authenticity of Meat and Meat Products

Meat consumption is one of the highest of the all food types across the globe. Meat and meat products belong to the most expensive foods and therefore fall into the category of frequently adulterated commodities. In 2013 news was made when beef was adulterated with horse meat which has some side effects for human consumption since horses are sometimes treated with medicines that are toxic to humans. Besides consumer safety there are ethical and religious concerns related to the consumption of some meat products. Therefore meat adulteration is a serious problem on a global scale. Customers are often deceived through the substitution of a high-quality meat with a less valuable kind or religious concern towards consumption of meat derived from a particular animal species. In order to prevent economic fraudulence and assure ethical concerns, different physical, chemical, or biochemical techniques are available. Nucleic acids or proteins among others are often used as markers. The most used techniques are gas chromatography, liquid chromatography, near-infrared spectroscopy, mid-infrared spectroscopy, matrix-assisted laser desorption/ionization mass spectrometry (MALDI-MS), electrophoresis with sodium dodecyl sulphate (SDS), capillary electrophoresis, isoelectric focusing, direct analysis in real time/time of flight mass spectrometry (DART/TOF-MS), enzyme-linked immunosorbent assay (ELISA), and polymerase chain reaction (PCR).

TABLE 3.2 Application of MALDI-MSI for the Analysis of Endogenous Molecule in Seeds and Grains

Sl.No	Sample Type	Technique	Analytes	Reference
1	Wheat seed	MALDI	Metabolites, amino acids, carbohydrates, oligosaccharides	Burrell et al. (2007)
2	Rice seed	MALDI	Lipids and other metabolites	Zaima et al. (2010)
3	Barley grain	MALDI	Lipids	Peukert et al. (2012)
4	Black rice seed	MALDI	Anthocyanins, lipids	
5	Cotton seed	MALDI	lipids	Horn et al. (2012)
6	*Camelina sativa* seed	MALDI	Lipids, triacylglycerols	Horn et al. (2013)
7	*Myristica malabarica* seed	MALDI	Alkaloids	Ifa et al. (2008)
8	Corn	MALDI	Metabolites (nucleosides, malic acid, lipids)	
9	*Brassica napus* seed	MALDI	Lipids	Woodfield et al. (2017)
10	*Arabidopsis thaliana* seed	MALDI	Lipids (phosphatidyl choline, phosphatidyl inositol)	Sturtevant et al. (2017)
11	Rice seed	MALDI	Finger print of phytochemicals	Kajiwara (2007)
12	Grape seed	MALDI	Proanthocyanin	
13	Rice	MALDI	Lyso phosphatidyl choline	Zaima et al. 2010

Buckley et al. (2009) present a robust method of analysing genus-specific collagen peptides extracted from bone fragments in 32 different mammal species, identifying a total of 92 peptide markers that could be used for species identification, for example, in processed food and animal feed. Cattle, pig, sheep, and chicken meat and processed bone meal samples were also analysed. To reduce intra-species variation, different individuals were sampled, for sheep (n = 10) and goat (n = 7), where different breeds were considered (two and three breeds, respectively). To simplify the analysis, on-tip C18 solid phase extraction was used instead of LC, before MALDI-TOF MS/MS.

3.4.3 Fruits and Vegetable Products

The MALDI equipment is gaining importance in the analysis of nutritional phytochemicals, the identification of geographical origin, and the identification of pathogenic bacterial strains causing various diseases in fruits and vegetable products. The phytochemicals have high variation in structure and concentration within and among plant species. The structural heterogeneity of polyphenols from cranberries, grape seed extracts, sorghum, and pomegranate have been studied. These were characterized by MALDI-TOF MS. It was demonstrated that large heterogeneity occurs in the degree of polymerization, intermolecular bonds, patterns of hydroxylation, and substitution with monosaccharides and gallic acid. MALDI-TOF was used to characterize the degree of polymerization (DP) and structure of proanthocyanidins (PA) oligomers in cranberries that are associated with the *in vitro* inhibition of adhesion of *E. coli* to uroepithelial cells. The methods detected oligomers up to a DP of 7 in the PA fraction that had the greatest anti-adhesion activity.

The identification of the geographical origin of food products is an international issue. Yoon et al. (2017) determined the geographical origin of kimchi in a traditional fermented-salted Korean side dish made of vegetables, typically prepared by fermenting Napa cabbage (*Brassica rapa* subsp. *pekinensis*) with various seasonings, e.g. red pepper, onion, garlic, and ginger. Kimchi is the most widely consumed item in Korea and has been consumed by Koreans for approximately 2,000 years. In this study, a rapid, reliable MALDI-TOF MS method, integrating multivariate analysis, was employed for the discrimination of fermented Korean and Chinese kimchi samples by geographical origin. Based on peaks extracted from the mass spectra on the basis of the *q*-value (<0.2), Chinese and Korean kimchi samples were discriminated within ten minutes.

Recently, microbial identification approaches have been developed using MALDI-TOF-MS as a fast, easy, and effective alternative for the identification of microbial isolates. Sindt et al. (2018) developed a novel workflow for strain typing using MALDI-TOF-MS coupled with PLS-DA and SIMCA classification with class-balanced Monte Carlo cross-validation. The study differentiated strains of *Xanthomonas axonopodis*, a bacterial plant pathogen that infects food and forage crops such as alfalfa, onion, and citrus.

3.5 CONCLUSION

MALDI-MS is a powerful tool that enables the simultaneous detection and identification of biomolecules in analytes. Nowadays the use of MALDI-MS technique in MALDI-imaging mass spectrometry is gaining more importance. MALDI-MS is used to visualize the spatial distribution of biomolecules without extraction, purification, separation, or labelling of biological samples. This technique has revealed the characteristic distribution of several biomolecules including proteins, peptides, amino acids, lipids, carbohydrates, and nucleotides. The versatility of this technique has opened a new frontier in several fields such as food, agriculture, medicine, pathology, and microbiology.

REFERENCES

Aparna, C. H., Madhavilatha, N., Supriya, P., and Gaurishankar, D. (2015). A review on matrix assisted laser desorption/ionisation mass spectroscopy. *Asian J. Pharm. Res.* 8(5): 28–33.

Asfandiarov, N. L., Pshenichnyuk, S. A., Fokin, A. I., Lukin, V. G., and Falko, V. S. (2002). Electron capture negative ion mass spectra of some typical matrix-assisted laser desorption/ionization matrices. *Rapid Commun. Mass Spectrom.* 16(18): 1760–1765.

Buckley, M., Collins, M., Thomas-Oates, J., and Wilson, J. C. (2009). Species identification by analysis of bone collagen using matrix-assisted laser desorption/ionisation time-of-flight mass spectrometry. *Rapid Commun. Mass Spectrum.* 23: 3843–3854.

Burrell, M., Earnshaw, C., and Clench, M. (2007). Imaging matrix assisted laser desorption ionization mass spectrometry: a technique to map plant metabolites within tissues at high spatial resolution. *J. Exp. Bot.* 58: 757–763.

Bjarnholt, N., Li, B., D'Alvise, J., and Janfelt, C. (2014). Mass spectrometry imaging of plant metabolites – principles and possibilities. *Nat. Prod. Rep.* 31: 818–837.

Dlott, D. D., Hambir, S., and Franken, J. (1998). The new wave in shock waves. *J. Phys. Chem. B.* 102(12): 2121–2130.

Dougherty, T. J., Kaufman, J. E., Goldfarb, A., Weishaupt, K. R., Boyle, D., Mittleman, A. et al. (1978). Photoradiation therapy for the treatment of malignant tumors. *Cancer Res.* 38(8): 2628–2635.

Dreisewerd, K., Schürenberg, M., Karas, M., and Hillenkamp, F. (1995). Influence of the laser intensity and spot size on the desorption of molecules and ions in matrix-assisted laser desorption/ionization with a uniform beam profile. *Int. J. Mass Spectrom. Ion Process.* 141(2): 127–148.

Hatakeyama, M. and Tachikawa, M. (2011). *Ab initio* quantum chemical study on the mechanism of *exceptional behavior* of lysine for ion yields in MALDI—role of vibrational entropic contribution in thermally averaged proton affinities. *J. Mass Spectrom.* 46(4): 376–382.

Horn, P. J., James, C. N., and Gidda, S. K. (2013). Identification of a new class of lipid droplet-associated proteins in plants. *Plant Physiol.* 162: 1926–1936.

Horn, P. J., Korte, A. R., and Neogi, P. B. (2012). Spatial mapping of lipids at cellular resolution in embryos of cotton. *Plant Cell* 24: 622–636.

Horneffer, V., Dreisewerd, K., Lüdemann, H.-C., Hillenkamp, F., Läge, M., and Strupat, K. (1999). Is the incorporation of analytes into matrix crystals a prerequisite for matrix-assisted laser desorption/ionization mass spectrometry? A study of five positional isomers of dihydroxybenzoic acid. *Int. J. Mass Spectrom.* 185–187: 859–870.

Ifa, D. R., Manicke, N. E., Dill, A. L., and Cooks, R. G. (2008). Latent fingerprint chemical imaging by mass spectrometry. *Science* 321: 805.

Ireland, J. J. and Wyatt, P. A. (1976). Acid-base properties of electronically excited states of organic molecules. *Adv. Phys. Org. Chem.* 12: 131–221.

Jaskolla, T. W. and Karas, M. (2011). Compelling evidence for lucky survivor and gas phase protonation: the unified MALDI analyte protonation mechanism. *J. Am. Soc. Mass Spectrom.* 22: 976–988.

Jaskolla, T. W., Karas, M., Roth, U., Steinert, K., Menzel, C., and Reihs, K. (2009). Comparison between vacuum sublimed matrices and conventional dried droplet preparation in MALDI-TOF mass spectrometry. *J. Am. Soc. Mass Spectrom.* 20: 1104–1114.

Karas, M., Bachmann, D., and Hillekamp, F. (1985). Influence of the wavelength in high-irradiance ultraviolet laser desorption mass spectrometry of organic molecules. *Anal. Chem.* 57(14): 2935–2939.

Karas, M. and Hillenkamp, F. (1988). Laser desorption ionization of proteins with molecular masses exceeding 10,000 daltons. *Anal. Chem.* 60: 2299–2301.

Lai, Yin-Hung and Wang, Yi-Sheng. (2017). Matrix-assisted laser desorption/ionization mass spectrometry: mechanistic studies and methods for improving the structural identification of carbohydrates. *Mass spectrom.* 6: 1–12.

Laiko, V. V., Baldwin, M. A., and Burlingame, A. L. (2000a). Atmospheric pressure matrix-assisted laser desorption/ionization mass spectrometry. *Anal. Chem.* 72: 652–657.

Laiko, V. V., Moyer, S. C., and Cotter, R. J. (2000b). Atmospheric pressure MALDI/ion trap mass spectrometry. *Anal. Chem.* 72: 5239–5243.

Leopold, J., Popkova, Y., Engel, K. M., and Schiller, J. (2018). Recent developments of useful MALDI matrices for the mass spectrometric characterization of lipids. *Biomolecules* 8: 173.

Liu, B. H., Charkin, O. P., Klemenko, N., Chen, C. W., and Wang, Y. S. (2010). Initial ionization reaction in matrix-assisted laser desorption/ionization. *J. Phys. Chem. B* 114: 10853–10859.

Molin, L., Seraglia, R., Czarnocki, Z., Maurin, J. K., Pluciński, F. A., and Traldi, P. (2012). On the primary ionization mechanism(s) in matrix-assisted laser desorption ionization. *J. Anal. Methods Chem.* 2012: 1–8.

Mowry, C. D. and Johnston, M. V. (1994). Internal energy of neutral molecules ejected by matrix-assisted laser desorption. *J. Phys. Chem.* 98: 1904–1909.

O'Rourke, M. B., Djordjevic, S. P., and Padula, M. P. (2018). The quest for improved reproducibility in MALDI mass spectrometry. *Mass Spectrom. Rev.* 37: 217–228.

Peukert, M., Matros, A., Lattanzio, G., Kaspar, S., Abadia, J., and Mock, H. P. (2012). Spatially resolved analysis of small molecules by matrix-assisted laser desorption/ionization mass spectrometric imaging (MALDI-MSI). *New Phytol.* 193: 806–815.

Schulz, E., Karas, M., Rosu, F., and Gabelica, V. (2006). Influence of the matrix on analyte fragmentation in atmospheric pressure MALDI. *J. Am. Soc. Mass Spectrom.* 17: 1005–1013.

Setz, P. D. and Knochenmuss, R. (2005). Exciton mobility and trapping in a MALDI matrix. *J. Phys. Chem. A* 109(18): 4030–4037.

Sindt, N. M., Robison, F., Brick, M. A., Schwartz, H. F., Heuberger, A. L., and Prenni, J. E. (2018). MALDI-TOF-MS with PLS modeling enables strain typing of the bacterial plant pathogen *Xanthomonas axonopodis*. *J. Am. Soc. Mass Spectrom.* 29: 413–421.

Sturtevant, D., Duenas, M. E., Lee, Y. J., and Chapman, K. D. (2017). Three-dimensional visualization of membrane phospholipid distributions in *Arabidopsis thaliana* seeds: a spatial perspective of molecular heterogeneity. *Biochim. Biophys. Acta* 1862: 268–281.

Szájli, E., Fehér, T., and Medzihradszky, K. F. (2008). Investigating the quantitative nature of MALDI-TOF MS. *Mol. Cell. Proteomics* 7: 2410–2418.

Tanaka, K., Waki, H., Ido, Y., Akita, S., Yoshida, Y., Yoshida, T., and Matsuo, T. (1988). Protein and polymer analyses up to m/z 100,000 by laser ionization time-of-flight mass spectrometry. *Rapid Commun. Mass Spectrom.* 2: 151–153.

Woodfield, H. K., Sturtevant, D., and Borisjuk, L. (2017). Spatial and temporal mapping of key lipid species in *Brassica napus* seeds. *Plant Physiol.* 173: 1998–2009.

Yoon, S. R., Kim, S. H., Lee, H. W., and Ha, J. H. (2017). A novel method to rapidly distinguish the geographical origin of traditional fermented salted vegetables by mass fingerprinting. *PLoS ONE* 12: 1–13.

Zaima, N., Goto-Inoue, N., Hayasaka, T., and Setou, M. (2010). Application of imaging mass spectrometry for the analysis of *Oryza sativa* rice. *Rapid Commun. Mass Spectrom.* 24: 2723–2729.

Zenobi, R. and Knochenmuss, R. (1998). Ion formation in MALDI mass spectrometry. *Mass Spectrom. Rev.* 17(5): 337–366.

Zhigilei, L. V., Leveugle, E., Garrison, B. J., Yingling, Y. G., and Zeifman, M. I. (2003). Computer simulations of laser ablation of molecular substrates. *Chem. Rev.* 103: 321–347.

Zhu, J. (1995). Analysis of conditional genetic effects and variance components in developmental genetics. *Genetics* 141(4): 1633–1639.

CHAPTER 4

MALDI IMS for Proteins and Biomarkers

Michelle L. Reyzer and Richard M. Caprioli

CONTENTS

4.1 INTRODUCTION

The introduction of matrix-assisted laser desorption/ionization (MALDI) [1–3] in the late 1980s ushered in a new era of biological research, allowing large, intact biomolecules, primarily proteins, to be analyzed by mass spectrometry (MS) for the first time. Building upon this innovation, for the past decade MALDI MS has been applied to the direct analysis of proteins and peptides from thin tissue sections [4–6]. This type of analysis is advantageous for several reasons, including minimal sample preparation requirements, relatively fast analysis, and the analysis of the active proteome complement that is present at a given location and time. One primary advantage of this technology is the ability to spatially map, or image, the distribution of any protein or peptide across the entire tissue section.

This ability to combine spatial information with high molecular specificity is unique to MS, and over the past decade MALDI imaging has flourished. The goal of this chapter is to briefly summarize the MALDI imaging mass spectrometry (IMS) experiment for proteins and biomarkers, to describe seminal research illustrating the strengths and versatility of the technique, and to elaborate on current advances, examples, and challenges relating to direct protein imaging by MALDI MS. While this technology is currently widely available and new applications and advances are continuing to be reported by many researchers, this chapter highlights examples from the author's laboratory.

4.2 MALDI IMS: A BRIEF HISTORY

As was first described in 1997 [4], the application of MALDI MS to the direct analysis of proteins in tissue sections is illustrated in Figure 4.1. The general procedure is to obtain a thin section of a sample of interest, apply matrix (which is necessary for the MALDI process), and then acquire mass spectra at discrete x-, y-coordinates over the entire sample section. Each individual mass spectrum contains signals corresponding to proteins (and other endogenous compounds, including peptides, lipids, and metabolites) present in the sample at a unique set of coordinates. The intensities of any given molecular species may then be plotted as a function of position on the sample surface.

In the imaging approach, the technology is a powerful discovery tool, highlighting areas of the tissue where certain proteins are abundant or absent, where proteins may colocalize with other proteins, and where proteins may localize in histologically well-defined areas. In addition, the signals obtained primarily represent the proteins as they are found in that region of tissue at that moment of time, including all the processing and/or posttranslational modifications they have undergone. Indeed, the first applications of this technology in the imaging mode showed localizations of proteins in well-defined and symmetric compartments of mouse brain tissue, as shown in Figure 4.2 [6]. Similarly, differential localizations of several proteins, including thymosin β4, β actin, and S100A4, were shown in different regions of a section of mouse xenograft glioblastoma tumor.

Another way of implementing this technology is in the profiling mode, which can simply be thought of as a low-resolution imaging experiment with high morphological specificity. In this mode, matrix is deposited in discrete locations upon the tissue surface corresponding to areas that are highly enriched in a cell population of interest (normal epithelial cells, areas of surrounding stroma, invasive cancerous cells, etc.). Mass spectra are acquired at each spot (so that the spectra are still linked to a location on the tissue surface), but the analysis is more typically a biostatistical analysis where protein profiles or signatures are compared between and among different regions, often with the goal of discovering biomarkers that can be used in diagnostic and/or prognostic analyses. In many

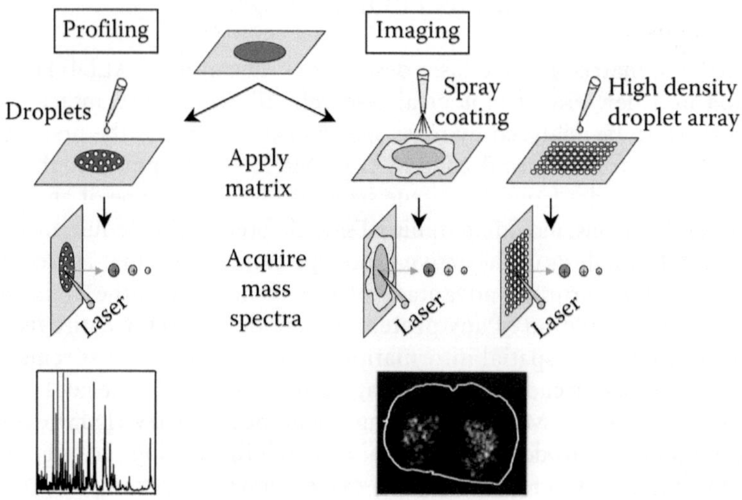

FIGURE 4.1 General procedure for MALDI MS imaging experiments in profiling (left side) and imaging (right side) modes.

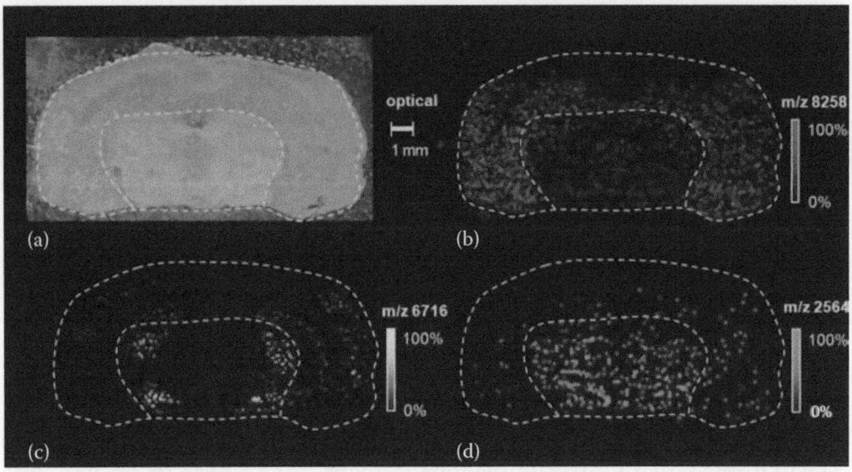

FIGURE 4.2 One of the first examples of mass spectrometric images of proteins in a mouse brain section. (a) Optical image of the tissue section mounted on a gold-coated plate. (b) *m/z* 8528 in the regions of the cerebral cortex and the hippocampus. (c) *m/z* 6716 in the regions of the substantia nigra and medial geniculate nucleus. (d) *m/z* 2564 in the midbrain. (Reproduced from Stoeckli, M. et al., *Nat. Med.*, 7, 493, 2001. With permission.)

cases, both imaging and profiling experiments are undertaken on the same samples, thus generating complementary data that, for example, visually show the location of statistically significant biomarkers.

The first applications of the profiling approach involved human lung cancer [7] and human brain cancer (glioma) [8, 9] samples. In both cases, groups of signals obtained from cancerous tissues were found to be significantly different from signals obtained from noncancerous tissues, and different statistical algorithms were utilized to successfully distinguish cancer from noncancer and to classify subtypes of tumors.

4.3 TECHNOLOGY DEVELOPMENTS: DATA ACQUISITION AND PROCESSING

In the decade since its inception, many significant advances have occurred to make the MALDI imaging process more widely available, more automated, and more robust. These improvements involved more sophisticated sample preparation methods as well as commercial instrumental and software improvements. In general, MALDI imaging may be performed on any mass spectrometer equipped with a MALDI source, but realistically some type of software improvements are necessary to facilitate data acquisition. Indeed, the first images were generated via manual control of the sample stage where data acquisition took ~1 min/pixel. Figure 4.3a shows an example of an "image" of rat pancreas tissue taken before any image generation or processing software was available. It shows a graphical representation of the location of the signal intensity at *m/z* 5802 (insulin) along a linear coordinate.

Thus, the first improvement was a software program written in-house (MS Image Tool) to automate instrument control, data sampling, and data processing [10]. This

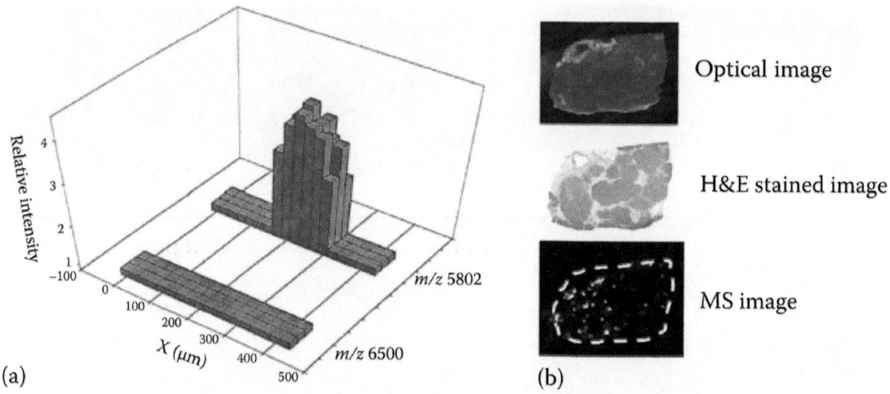

FIGURE 4.3 Technological developments have facilitated the development and processing of MALDI image data. (a) Rudimentary "image" of an area of rat pancreas tissue, 450 × 75 µm. The reconstructed ion image at *m*/*z* 5802 shows the presence of insulin in an islet while the image at *m*/*z* 6500 serves as a control. (Reproduced from Caprioli, R.M. et al., *Anal. Chem.*, 69, 4751, 1997. With permission.) (b) Software developments allowed this image of human normal pancreas tissue (adjacent to tumor) to be obtained. From top to bottom: optical image of the tissue section, H&E stained section, and MALDI MS image of insulin (*m*/*z* 5808) visualized with Biomap software obtained at 150 × 150 µm resolution. Insulin is clearly visualized within many islet cells.

improved data acquisition times to ~1 s/pixel or better. Currently, software to automate image acquisition and perform some image processing is commercially available from several MS instrument vendors, including FlexImaging from Bruker Daltonics, ImageQuest from Thermo Scientific, MassLynx from Waters, and oMALDI server [11] and TissueView from Applied Biosystems/MDS Sciex. These tools allow the user to specify instrumental parameters such as spatial resolution, laser shots per spot, and pixel coordinates, as well as providing some data processing and image generation capabilities. Along with higher frequency lasers and faster electronics, these software packages allow for imaging times on the order of 10–100 ms/pixel (depending on desired resolution and sensitivity). An example of an image acquired using current technology of similar tissue (human pancreas) is shown in Figure 4.3b. Clearly, advances in the imaging process are visible, as the expression of insulin (*m*/*z* 5808) is more accurately visualized as part of the tissue section.

In addition, several software programs are freely available to assist with the imaging process, including BioMap from Novartis [12] (which is an image process- ing program for use on any data set that has been transformed into the Analyze file format), *CreateTarget* and *Analyze This!* [13] (which work with Bruker instruments to create custom geometry files and to export data into the Analyze format for further analysis with BioMap), and Axima2Analyze (converts data from Shimadzu Axima QIT instruments to Analyze format). These software packages primarily assist with image generation and data processing, and are currently available for download from the Mass Spectrometry Imaging interest group website (www. maldi-msi.org).

In the interest of generating a "universal" tool that can be used to analyze MS data acquired from any one of the various MS instruments commonly used for peptide and protein analysis (often with their own proprietary data file structure), the Human Proteome Organization (HUPO) has proposed creating a standard format for MS data, called mzML (building on a previous version, mzXML) [14]. A new imaging software package has been developed to take advantage of the XML format, MITICS [15], which serves to both control image data acquisition on Applied Biosystems MALDI-TOF Voyager series instruments (MITICS Control) and provide for data processing and image reconstruction in the XML format (MITICS Image). An extension of the standard MS data format, called i-mzML (www.i-mzml. org), for imaging data has also been proposed. This trend toward making all imaging data available in one standard format will assist in image evaluation and validation between and among different acquisition platforms and should allow further development of the technology to progress more rapidly.

4.4 TECHNOLOGY DEVELOPMENTS: AUTOMATION

In addition to the increased availability of image-specific software packages, there are a number of tools now commercially available that are designed to automate various aspects of sample preparation. The ultimate goal for imaging applications is to acquire images that accurately show the distribution of proteins while maintaining spatial integrity. The requisite addition of a matrix to the tissue section, typically as a solution, introduces the possibility of analyte delocalization as analytes may be solubilized by the matrix solution and migrate over the surface of the tissue section. Two instruments are now commercially available that enable automated spray-coating of matrix on tissue sections. Bruker Daltonics manufactures the ImagePrep system, which nebulizes small amounts of matrix solution and allows the "mist" to fall onto a tissue section mounted on a glass slide. By monitoring the transmission of light through the glass slide, the instrument can estimate the relative "wetness" of the matrix coating, and optimize the timing for future coats of matrix. The TM-Sprayer by LEAP Technologies [16] utilizes a spray nozzle to create a fine mist of matrix, which is deposited on the sample plate as it is moved in an x, y pattern under the spray nozzle. One recent report also describes the use of a commercial desktop inkjet printer for automated matrix deposition [17]. This approach provides a less expensive alternative to the commercially available robotic sprayers and accommodates the possibility of using multiple solvents.

The ultimate goal of these robotic instruments, as with manual spray-coating, is to generate a thin, homogeneous coating of matrix crystals over the tissue surface while maintaining the spatial integrity of the solubilized proteins. Automating the process should make it more reproducible and thus improve the quality of the images produced.

4.5 TECHNOLOGY DEVELOPMENTS: SAMPLE PREPARATION

Apart from automating the matrix application process, it is critical to evaluate the resulting matrix crystals and coating for analyte extraction, localization, and effect on tissue architecture. Several reports have shown that even for standard analytes mixed with matrices, there is an uneven distribution of the analyte within the MALDI crystals and that sample preparation influences the resulting distribution [18–20]. Thus, other active areas of research are focused on optimizing matrix application and sample preparation protocols.

For example, one way to avoid solvent-induced analyte delocalization is to apply matrix without any solvent. Two groups have explored this approach, one using sublimation to apply matrix to tissue sections [21], and the other using finely ground powdered matrix applied through a sieve [22]. Both of these approaches were evaluated on mouse brain tissue and were found to produce homogeneous coatings consisting of fine microcrystals (<20 μm). Relatively high-resolution images (~30–100 μm) were acquired from various phospholipids present in the brain tissues illustrating high degrees of analyte localization within various substructures. An example of the DHB crystals obtained after dry-coating DHB on brain tissue is shown in Figure 4.4a, along with representative images of phospholipids obtained from spray-coating and dry-coating mouse brain sections (Figure 4.4b). While these results are very promising for lipids, the applicability of these techniques to larger, more polar analytes, including peptides and proteins, is yet to be fully examined.

Another approach to matrix application is to robotically deposit small (pL to nL) droplets of matrix on discrete areas of the sample surface. When the droplets are deposited as arrays over the entire tissue surface, mass spectra can be acquired at each matrix spot and reconstructed into an image. In this case, potential analyte delocalization is limited to the area under the matrix spot, typically ~100–250 μm with commercially available spotters (e.g., Portrait 630 by Labcyte [23], ChIP by Shimadzu, TM iD by LEAP Technologies). This approach is fundamentally a trade-off between spatial integrity and resolution, because what limits resolution in most cases is the diameter of the laser beam, which is typically much smaller than the diameter of the matrix spot (on the order of 30–50 μm). Samples prepared by spray-coating, either manually or robotically, may be imaged at the diameter of the laser beam, but may not have maintained analyte localization throughout the spray process.

Another goal of matrix application, important to diagnostic samples, is the preservation of tissue integrity for subsequent histological analysis. Classical histological

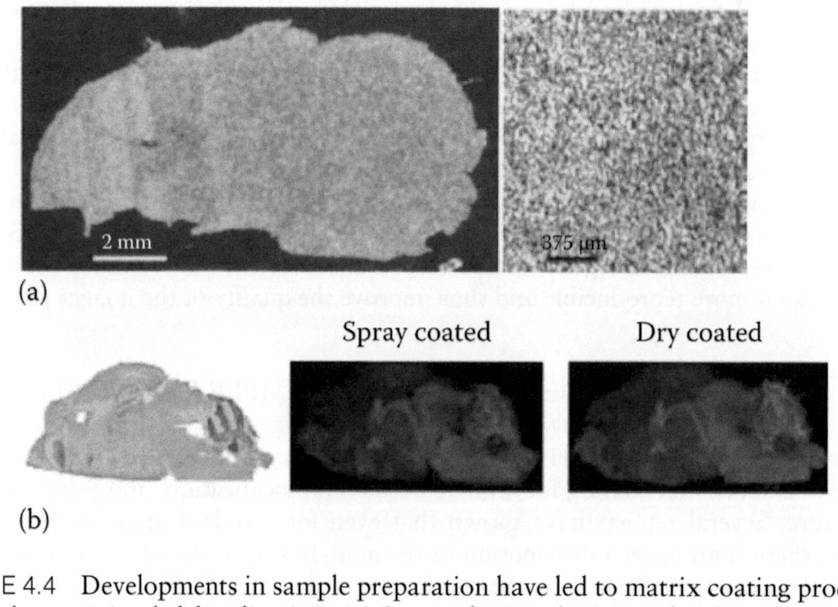

FIGURE 4.4 Developments in sample preparation have led to matrix coating procedures that induce minimal delocalization. (a) Scanned optical image of a dry-coated sagittal mouse brain section on a gold-coated MALDI plate, with a 4× magnification of the DHB layer shown on the right. (b) Comparison of phospholipid images obtained from a sagittal mouse brain either spray-coated (middle) or dry-coated (right) with DHB. (Adapted from Puolitaival, S.M. et al., *J. Am. Soc. Mass Spectrom.*, 19, 882, 2008. With permission.)

techniques usually involve immersing the tissue in a solution, such as formalin or ethanol, to fix the proteins while maintaining cellular structure. Formalin works by crosslinking proteins, and thus is not readily compatible with MALDI MS analysis, although some work has been done with samples fixed in this way using alternative approaches [24]. (This will be discussed in more detail in the application section.) Implementing solvent rinses as a means to improve peptide and protein sensitivity for MALDI IMS presupposes that the solvent will rinse away compounds on the tissue surface, primarily salts and lipids, that compete for protons or otherwise inhibit protein cocrystallization and/or ionization. However, any time the tissue is immersed in a liquid, the possibility exists first that analytes of interest may be soluble in the rinsing solvent and thus be extracted and lost, and second that soluble analytes may be extracted and redeposited elsewhere on the tissue surface, thus inducing delocalization. Methodology development must incorporate a full investigation of these effects prior to the implementation of any washing protocol, and it must be kept in mind that the washes may behave differently on different tissue surfaces (i.e., liver vs. brain vs. tumor). Nonetheless, different rinses and fixation protocols have been described by several groups and by and large have been shown to improve MALDI sensitivity for peptides and proteins while maintaining tissue architecture and preserving the tissues over time [25–29].

First reported as a method to improve the crystal coverage and crystal formation of α-cyano-4-hydroxycinnamic acid (CHCA) on tissues, largely due to the removal of surface salts and lipids [29], further studies have systematically investigated the tissue fixation properties of various organic solvent rinses [26, 27]. Lemaire et al. [27] developed a tissue-washing procedure based on organic solvents typically used for lipid extraction. Using rat brain sections as the standard tissue, they compared MALDI signals in the protein range with and without rinsing the tissue with different organic solvents, including chloroform, hexane, acetone, toluene, and xylene. They found a 44% increase in the number of signals detected ($m/z > 3000$) in tissue rinsed with xylene and a 34% increase in tissue rinsed with chloroform compared to control (untreated) tissue, as well as an increase in overall signal intensity. A decrease in the standard deviation was observed for all solvents tested. Tissue cellular structure was determined with optical microscopy and immunohistochemistry to be intact after rinsing with no notable delocalization of proteins.

Imaging experiments in both the peptide and protein range were undertaken with rat brain sections at least six months old prior to analysis [27]. Two adjacent sections were obtained; one was left untreated and the other was rinsed with chloroform. Sinapinic acid was applied as the matrix for protein imaging. For older tissues it was difficult to recognize brain structures without any treatment, but after rinsing the structure became apparent. The signal obtained for a relatively high mass protein at m/z 14,306 showed an increase in intensity after chloroform rinsing compared to the untreated section. Visualizing the spatial localization of the protein was thus enhanced after chloroform rinsing. The authors concluded that treatments with these organic solvents improved the detection and sensitivity of proteins via lipid removal without generating any peptide or protein delocalization. They also noted however, that there was not a reduction of salt adducts as has been shown with alcohol rinses.

Seeley et al. also systematically evaluated a series of organic solvents for the effects on protein signals from tissue using rat liver tissue as a model [26]. Alcohol treatments (ethanol, methanol, isopropanol) were evaluated in addition to lipid-removing solvents (chloroform, xylene, toluene, hexane). In this case the goal was to determine the overall effect on protein signal after rinsing (number of signals, signal intensity) as well as tissue preservation, that is, how long after sectioning/rinsing/matrix application could the tissue be analyzed with minimal disruption to the signals. Similar to Lemaire, they found increases in both the number of protein signals and the signal reproducibility after most washes (except acetone) compared to the untreated control. They also reported that overall signal intensity and quality were maintained with no significant deterioration after washing with isopropanol for up to one week. Ethanol and isopropanol were also found to leave the tissues virtually indistinguishable from untreated tissues after histological staining, thus indicating no disruption of tissue architecture [26].

Pathologists use staining to enhance the contrast between cellular regions on tissue sections to diagnose clinical specimens, with hematoxylin and eosin (H&E) being a common staining protocol. Combining this histological information with MALDI analyses of protein expression is quite powerful for biomarker discovery, because the cellular location of specific biomarkers can be assessed. However, this is typically accomplished using serial sections of a given tissue specimen because common staining protocols, such as H&E, have been found to compromise the MALDI signal [30]. Alternative staining protocols were thus evaluated in order to develop a protocol that could be used both for histological assessment and for MALDI MS. Toluidine blue, nuclear fast red, methylene blue, Terry's polychrome, and cresyl violet were applied to mouse liver sections in either water- or ethanol-based solvents [28]. Excess stain was removed with ethanol washing, and the resulting stained tissues were evaluated both for optical detection of cellular features and for the resulting MALDI spectra after matrix application. In this case, MALDI signals from methylene blue and cresyl violet were similar to control (ethanol-washed) tissue, in terms of number of signals and signal intensity. It was also shown that cresyl violet offered high nuclear to cytoplasm contrast necessary for adequate histological analysis of cancer tissue [28]. This analysis allows for MALDI IMS and histological analysis to be performed on the same tissue section instead of serial sections, which can eliminate any errors associated with sectioning and is valuable for very small clinical diagnostic specimens, such as needle biopsies.

Agar et al. also investigated histology-compatible tissue preparation protocols for MALDI IMS [25]. In this case, they modified several established immunohistochemistry protocols to include dissolved MALDI matrices in the solvent rinses, and they examined the resulting tissue architecture and MALDI protein signals using mouse brain tissue as a model. They found an ethanol/methanol/acetonitrile/0.1% trifluoroacetic acid (in water) mix in a 2:2:1:1 ratio containing 25–30 mg/mL of matrix resulted in the best MALDI signal (in terms of number of signals and S/N of the resulting signals) while minimizing diffusion of proteins and tissue deformation. Again, this protocol would allow one single tissue section to be stained, visually examined for histological features, and then subjected to MALDI IMS analysis where the resulting protein images could be directly correlated with tissue architecture.

4.6 EXAMPLES AND APPLICATIONS

MALDI MS offers the ability to discretely analyze the proteomic composition of many well-defined substructures in various tissues, without the need for dissection, extraction, or homogenization. It has thus been used as a tool for the study of normal development [31] and of many varied pathological processes, used in a targeted or nontargeted approach. A wide variety of tissues, both animal and human, including brain [6, 12, 25, 28, 31, 32], liver [26, 28], kidney [33, 34], eye [35], pituitary [36], pancreas [4], prostate [37], colon [5], and breast [38], have been analyzed using this methodology. Analyses of whole-body animal sections, rat or mouse, have also been described [39–41]. Pathological conditions, such as cancer [6–9, 24, 26, 38, 42–47], Parkinson's disease [48, 49], Alzheimer's disease [12], diabetes, tuberculosis, pre-term labor, and heart disorders, have been examined. These lists are not meant to be inclusive, and certainly with time the kinds of tissues and conditions examined will expand. Rather, this chapter aims to highlight several applications that illustrate the unique capabilities and versatility of the MALDI IMS experiment.

4.6.1 Applications to Development and Disorders

The study of Parkinson's disease is often done in animal models where one side of the brain is lesioned, generally via administration of a chemical, such as 1-methyl-4- phenyl-1,2,3,6-tetrahydropyridine (MPTP) or 6-hydroxydopamine (6-OHDA), that disrupts the dopaminergic pathway. Differences in protein expression between the lesioned and unlesioned sides of the brain have been analyzed via MALDI MS. One study revealed significantly decreased levels of Purkinje cell protein 4 (PCP-4 or PEP-19) in the striatum of MPTP-lesioned mice compared to control [48]. This result was in agreement with data acquired from a variety of other analytical techniques, including liquid chromatography–tandem mass spectrometry, high-resolution MS, and *in situ* hybridization. In a more recent study, the same group used a targeted approach to evaluate the distribution and relative quantitation of the immunophilin FKBP-12 in the striatum of 6-OHDA-lesioned rats [49]. MALDI MS profiling revealed increased levels of intact FKBP-12 (*m/z* 11,791) in the dorsal and middle parts of the dopamine-depleted striatum, but no difference in the level of FKBP-12 in the ventral part compared to the striatum in the intact hemisphere of the brain. These results were also supported by data from Western blotting and two-dimensional (2D) gel electrophoresis.

MALDI IMS was recently utilized to examine embryo implantation in mice [50]. A comprehensive understanding of the spatial and temporal changes that occur in the uterine environment prior to and during embryo implantation has yet to be realized. Successful implantations are necessary for successful pregnancies, so understanding the reasons for implantation failure has important ramifications for understanding and potentially treating some causes of infertility. However, because the biochemical events surrounding implantation are quite complex, with the roles and expressions of a large number of gene products remaining poorly understood during this process, this is an analytically challenging task. In addition, these implantation sites are approximately 2 mm in diameter, necessitating an analytical technique capable of high-resolution analysis. MALDI IMS allowed the authors to analyze hundreds of proteins involved in various biochemical processes during the peri-implantation period, and thus provide unique proteomic snapshots of implantation and interimplantation sites.

Implantation and interimplantation sites were excised from dissected mouse uteri on several days postimplantation and prepared for MALDI MS analysis. Profiling experiments were undertaken where matrix was deposited on discrete morphological areas of the tissue, and proteomic signatures were recorded. It was determined that 50 signals were significantly changed in the uterus due to the presence and proximity of the embryo [50]. Imaging experiments were subsequently undertaken, and the resulting images for several of the statistically significant proteins, including calcyclin and ubiquitin, are shown in Figure 4.5. As shown, both the location and expression of the proteins change

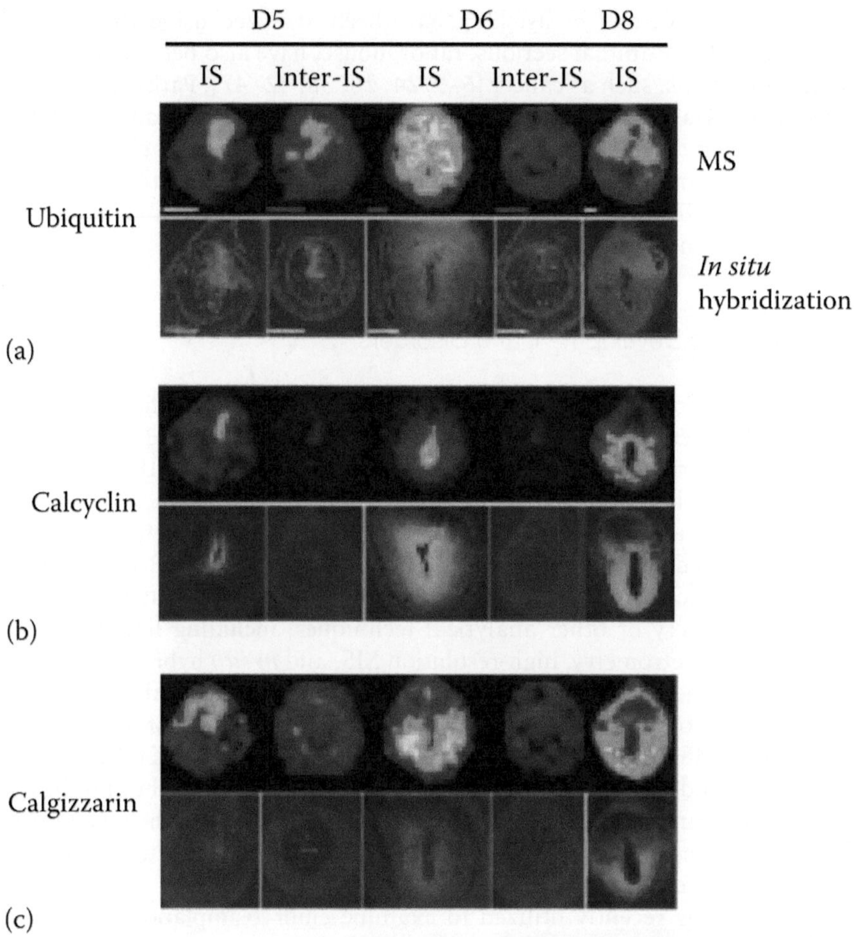

FIGURE 4.5 Imaging MS shows spatial localization for selected proteins of interest on days 5, 6, and 8 of pregnancy. Spatial localization of (a) ubiquitin, (b) calcyclin, and (c) calgizzarin by imaging MS (upper panels) and respective mRNAs by *in situ* hybridization (lower panels) in the implantation site (IS) or interimplantation site (inter-IS) of the mouse uterus. Bar, 700 μm. (Adapted from Burnum, K.E. et al., *Endocrinology*, 149, 3274, 2008. With permission.)

as a result of time (days 5, 6, or 8 after implantation). The imaging results are in good agreement with the profiling results and mRNA localizations as determined by *in situ* hybridization (Figure 4.5).

MALDI IMS was further used to examine the proteomic profiles of wild type and *Pla2g4a* null mice, which have known defects in implantation and experience implantation delays. The results of these studies suggest that proteome signatures are different between wild type and *Pla2g4a* null mice, which may help distinguish which proteins are critical for on-time implantation, which leads to more positive pregnancy outcomes.

The evolution of protein expression during normal mouse prostate development was examined with MALDI IMS technology [37]. In this case, protein expression and spatial localization was followed as a function of lobe specificity and age: during organ development (1–5 weeks), at sexual maturation (6 weeks), and as adult (10, 15, or 40 weeks). In addition, the protein signatures from normal development were compared to 15-week prostate tumors. Several protein signals were found to be uniquely expressed during the very early time points, such as *m/z* 4937 and *m/z* 4965 (consistent with the molecular weights of thymosin β 10 and 4, respectively), suggesting these proteins may play a role in early prostate development. In contrast, signals identified as probasin and spermine-binding protein (SBP) were increasingly detected as the prostate sexually matured. Most proteins showing this trend were absent in the tumors, indicating that normal protein synthesis was interrupted as the cells dedifferentiated. In addition, numerous signals were found to be lobe-specific.

One of the strengths of this technique is the ability to identify posttranslationally modified proteins by following a change in mass. In this study, the authors reported a change in the proportion of α-*N*-acetylated (*m/z* 17,882) and nonacetylated (*m/z* 17,840) cyclophilin A (Cyp A) between the normal prostate and prostate tumors. (The identity of Cyp A was confirmed by HPLC separation, MALDI MS peptide fingerprinting, and MS/MS analysis.) These results were confirmed by Western blotting and immunohistochemical analyses [37].

4.6.2 Three-Dimensional Imaging

The ability to localize biomarkers in three dimensions using MALDI imaging MS is an emerging approach for the analysis of protein profiles in organs. The initial proof-of-principle study optically registered 264 coronal mouse brain sections (20 μm thick) spanning the area of the corpus callosum to a stereotaxic coordinate system to produce a three-dimensional (3D) volume [32]. MALDI protein profiles were obtained on 10 of the 264 sections, which were approximately 400–500 μm apart, and the protein data inserted into the 3D volume. Thus, the distribution of any protein signal could be visualized not only within each coronal brain section but also over the entire 3D structure. Additional studies of 3D imaging have recently been reported for the rat brain model. The colocalization of not only proteins but also peptides or small endogenous or exogenous compounds were assessed in a 3D space.

One of the strengths of the MALDI MS technique for biomarker discovery is that many different types of compounds can be analyzed on a single analytical platform. This unique feature was utilized in a study examining the distribution of proteins and peptides in the same 3D volume from the substantia nigra and interpeduncular nucleus [51]. For this analysis, two series of adjacent sections were cut every 200 μm throughout the rostrocaudal axis of the rat midbrain. One section was coated with sinapinic acid as matrix for protein analysis and the other with DHB for peptide analysis. The matrix was spotted using a robotic matrix dispenser in an array with 200 μm spacing in both the *x*- and *y*-directions. As the sections were acquired every 200 μm (the *z*-direction), a single voxel consisted of the intensity of any *m/z* signal at one 200 × 200 × 200 μm point in 3D space. The small distance (12 μm) between adjacent sections taken for protein and peptide analyses was determined to be roughly the size of one medium-sized neuron, and thus negligible in terms of shifting the morphology in the regions, compared to the large 200 μm distance between points in 3D space. Similar to the initial experiment, a 3D volume of the analyzed brain area was constructed based on registering anatomical features, and the protein and peptide data were inserted into the 3D volume. Figure 4.6 illustrates the process from intrasection registration of the MALDI images with the matrix spots (Figure 4.6a), to creation of the 3D volume (Figure 4.6b), to the visualization of unique

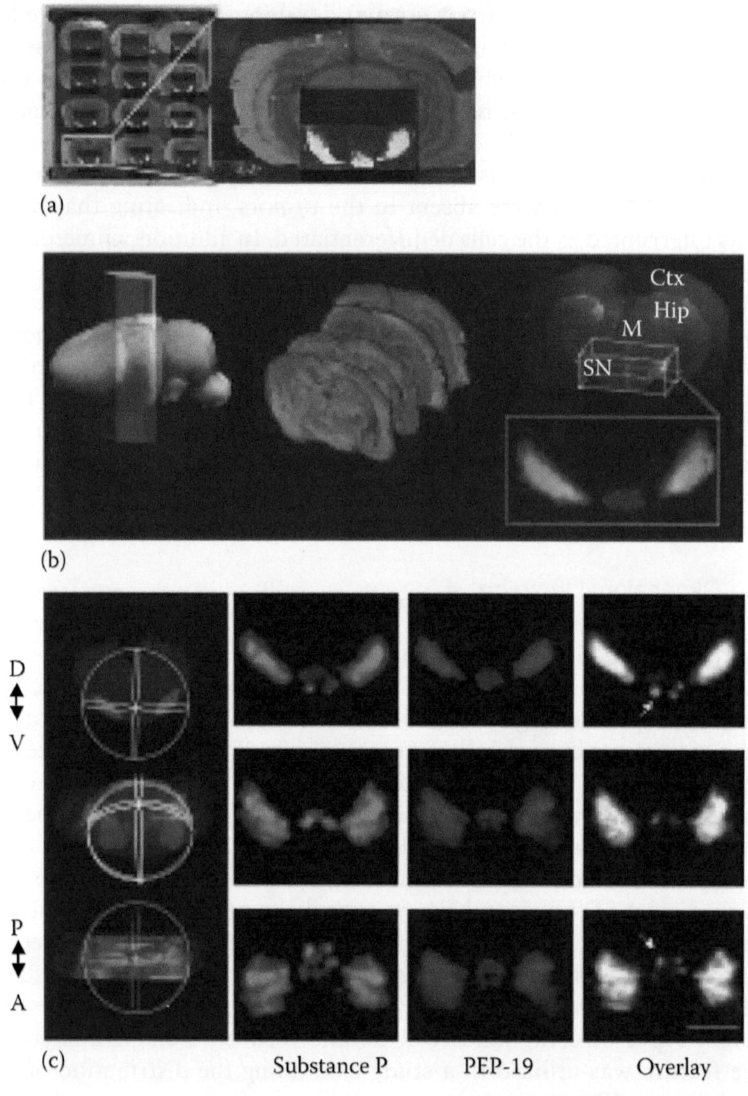

FIGURE 4.6 3D volume reconstructions of one peptide and one protein as detected by MALDI IMS of the rat ventral midbrain. (a) Intrasection registration of pixelated IMS images is performed so that each individual pixel matches its matrix deposit. (b) The intersection registration uses cresyl violet stained images to create a *z*-stack for alignment using Amira software. The 3D surface reconstructions of cortex (Ctx), hippocampus (Hip), midbrain (M), and substantia nigra (SN) visualize the alignment. The box outlines the volume analyzed by MALDI IMS. (c) MALDI IMS images of substance P (*m/z* 1347.8 [M + H]+) and PEP-19 (*m/z* 6717 [M + H]+) were registered using the same factors obtained for intra- and intersection registration. Virtual *z*-stacks were created in Image J software and 3D volume reconstructions of substance P and PEP-19 were rotated for frontal and dorsal views (top and bottom rows, respectively). Overlap of substance P and PEP-19 localization was assessed by combining *z*-stacks. Differential localization was observed in the interpeduncular nucleus. Anterior (A), posterior (P), ventral (V), dorsal (D). Scale bars, 2 mm. (Reproduced from Andersson, M. et al., *Nat. Methods*, 5, 101, 2008. With permission.)

peptides and proteins in the same 3D space (Figure 4.6c). The results showed significant overlap for the peptide substance P (*m/z* 1347.8, shown in green) and the protein PEP-19 (*m/z* 6717, shown in purple) in the substantia nigra, while differential localization was observed in the interpeduncular nucleus [51].

Extending the visualization of proteins in three dimensions, another recent study attempted to integrate the 3D representation of MALDI MS protein data with 3D data acquired from another imaging modality, namely, magnetic resonance imaging [44]. This study utilized a mouse with a brain tumor, where the whole mouse head was subjected to MRI imaging on a 7T magnet, after which the mouse was sacrificed and frozen. Sequential cryosectioning for subsequent MALDI analysis was then performed, with high-resolution digital images acquired for each blockface section. In this case, the 3D volume recreated from the blockface images was coregistered to the magnetic resonance volume of the mouse's head. Due to the fact that the brain is surrounded by bone, minimal deformations occurred to the brain during processing, so a relatively simple registration was required. Ion volumes were created for two proteins, the astrocytic phosphoprotein Pea 15 (*m/z* 15,035) and fatty acid-binding protein 5 (Fabp5, *m/z* 15,076) obtained from the MALDI data and rendered against the MRI data. These results showed the feasibility of correlating *in vivo* anatomical imaging with spatially resolved *ex vivo* molecular imaging. This is quite promising for biomarker discovery, as functional information from *in vivo* studies, such as tumor angiogenesis, can be directly correlated with molecular signatures from these areas.

4.6.3 Markers of Toxicity and Infection

MALDI MS has been utilized as part of a new strategy to investigate potential markers of toxicity. The kidney is often the major site of drug toxicity because it serves to filter and excrete exogenous compounds from the body. Nephrotoxicity is usually reversible if detected early so the toxicant can be discontinued, however early detection is not always straightforward [52]. Such early detection would have benefits in both the clinical setting as well as drug discovery. In one study, the known nephrotoxicant gentamicin was administered to rats for seven consecutive days [33]. Sections were acquired from kidneys from both control and gentamicin-treated rats, and proteomic profiles and images were acquired. Differential protein expression was observed within the normal kidney that distinguished the substructures of the kidney, with unique signals expressed in the cortex, medulla, and papilla.

The gentamicin-treated kidneys showed differential expression of a number of signals compared to controls, mostly in the cortex, consistent with the known action of gentamicin. This correlated well with H&E-stained sections, where the treated sections showed clear epithelial degeneration and regeneration in the proximal tubules. The most significant signal at *m/z* 12,959 was highly upregulated in the treated cortex but not detectable in the control (Figure 4.7). This signal was identified as transthyretin, a blood transporter protein. This result was confirmed by Western blotting and immunohistochemical staining. Several other signals were found to be differentially expressed in treated vs. control kidneys, and the use of IMS enabled the localization of the affected proteins. For example, a signal at *m/z* 8242 was found to be present throughout the cortex and medulla in the control kidney, but depleted only in the cortex after gentamicin treatment. Further studies are required to discover whether transthyretin is a general marker for kidney toxicity, a marker for proximal tubule damage, or is specific for gentamicin- or

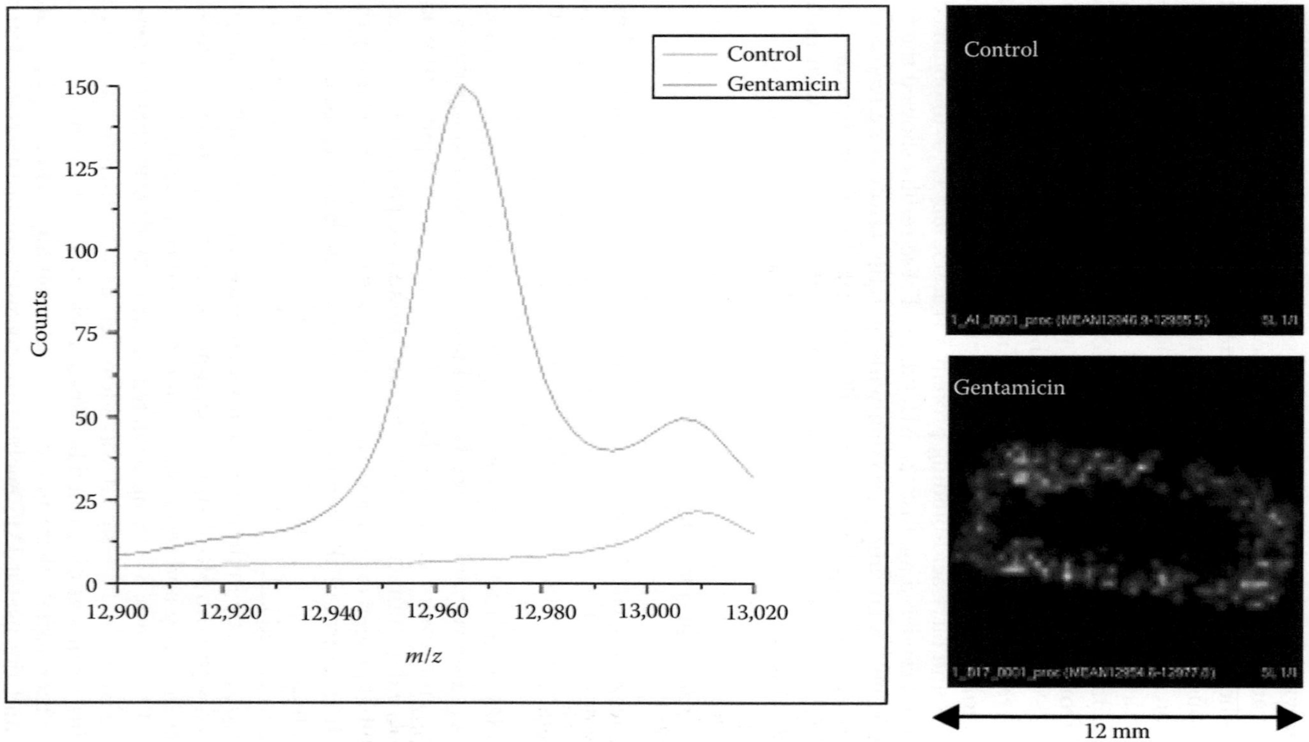

FIGURE 4.7 Differential imaging analysis of a control vs. a gentamicin-treated rat kidney section. Average spectrum (cortex) and intensity distribution of m/z 12,959 on a control and a treated kidney section. (Adapted from Meistermann, H. et al., *Mol. Cell. Proteomics*, 5, 1876, 2006. With permission.)

aminoglycoside-induced toxicity. Nonetheless, this result is promising and provides additional information not obtainable by conventional 2D gel electrophoresis analysis of kidney homogenates [53] or plasma samples [54].

Understanding the molecular events underlying the immune response to infection may lead to the design of better antimicrobial agents, a task that grows more urgent as more bacteria become resistant to current treatments. MALDI MS was utilized to examine host factors that limit microbial growth in abscesses in a mouse model of *Staphylococcus aureus* infection [34]. An initial imaging experiment comparing kidneys from control mice and mice infected with *S. aureus* revealed high signals from a protein at *m/z* 10,165 that was localized to the abscesses. This signal was subsequently identified as S100A8 (calgranulin A). Other imaging experiments also revealed a signal at *m/z* 12,976 that colocalized with S100A8 in the abscesses of infected kidneys. This protein was identified as S100A9 (calgranulin B). These two proteins make a functional heterodimer, calprotectin, which is a calcium-binding protein known to inhibit the growth of many bacterial pathogens. It was further shown that the proteins are only present in the abscesses in immunocompetent (neutrophil-replete) mice. A more thorough investigation of how calprotectin exerts its antibacterial activity determined that calprotectin likely starves the bacteria of the metal ions (Ca^{2+}, Mn^{2+}, Zn^{2+}) that *S. aureus* needs to survive [34].

4.7 NEW APPROACHES

A limitation of protein technologies is the relatively small subset of proteins that are readily amenable to detection (typically small proteins <50,000 Da, soluble proteins, and relatively abundant proteins). One approach to making this technology more suitable for other types of compounds, such as membrane proteins or mRNA, is the use of reporter tags. This strategy can be used to analyze compounds not generally amenable to MALDI, to multiplex the analysis of such compounds [55] (i.e., label several compounds with reporters of different mass so they may all be analyzed in one experiment), or to amplify the signal obtained from a low-abundance analyte [56].

A recent report describes the use of MALDI profiling (untargeted) followed by a targeted imaging approach, to determine biomarkers for ovarian cancer [46]. In this study 25 late-stage ovary carcinomas and 23 benign ovaries were analyzed using MALDI MS. One putative biomarker was found in ~80% of the ovarian cancer samples at *m/z* 9744. This signal was subsequently identified as a fragment of the 11S proteasome activator complex, Reg-alpha. Western blotting and immunocytochemical studies were carried out to validate the specific presence of this protein in cancer samples compared to benign ovaries. Finally, a targeted imaging experiment was carried out, with a primary antibody to Reg-alpha applied to the carcinoma tissue. The primary antibody was associated with a secondary antibody that had been modified with a photocleavable linker and a reporter peptide. MALDI analysis of this tissue section detected the reporter peptide at the sites that were occupied by the anti-Reg-alpha primary antibody. These results correlated well with the untargeted MALDI imaging experiment showing the localization of the fragment at *m/z* 9744. Taken together, these findings support the 11S proteasome activator complex as a potential biomarker of ovarian cancer.

Another approach is to digest proteins directly on tissue sections and then analyze the resulting peptides. This technique can make peptides from insoluble membrane proteins or high molecular weight, low-abundance proteins available for analysis, and it can provide a built-in validation, as all peptides from the same protein should have

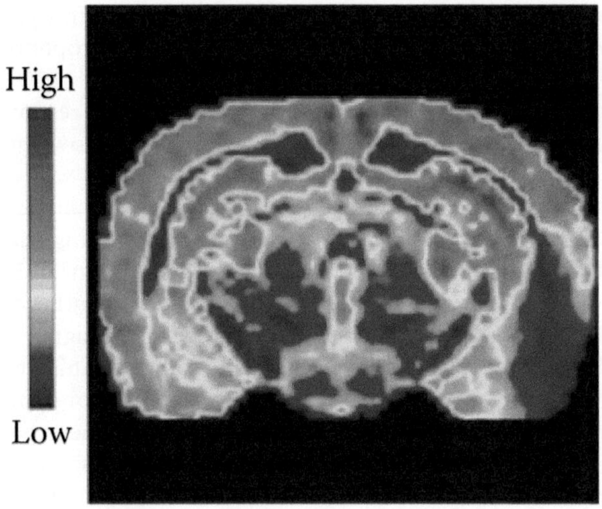

FIGURE 4.8 On-tissue digestion aids the detection of high-molecular-weight species. Distribution of ion detected at *m/z* 1496.75 corresponding to a tryptic peptide of the 71 kDa protein, synapsin I. (Reproduced from Groseclose, M.R. et al., *J. Mass Spectrom.*, 42, 254, 2007. With permission.)

the same distribution. This was recently demonstrated for proteins in rat brain tissue [57]. For example, tryptic fragments from the higher molecular weight proteins actin (41 kDa), tubulin (55 kDa), and synapsin-1 (74 kDa) were all identified using *in situ* tryptic digestion on a single tissue section. The distribution of one tryptic peptide from synapsin-1 (*m/z* 1496.75) is shown in Figure 4.8. In addition, this approach can also be a means to profile formalin-fixed tissue specimens whose native proteins have been crosslinked by the formalin fixation. This approach has tremendous potential, given the vast stores of archived samples in tissue banks that have been fixed in this manner. There are several reports describing the use of immunohistochemical antigen retrieval methods for more general proteomic profiling studies [58–60]. These results suggest that it is possible to extract proteomic identifications that are similar to those obtained from fresh frozen tissues.

One example of using on-tissue digestion for formalin-fixed tissue analysis was recently reported [24]. In this application, tissue microarrays (TMA) were created by embedding formalin-fixed needle core biopsies from lung tumor patients in paraffin blocks. Analysis of TMAs allows for hundreds or thousands of samples to be processed together in a high-throughput approach. Analyzing TMAs by MALDI IMS offers the additional advantage of molecular and spatial specificity. The TMA described in this paper contained 100 duplicate needle core biopsies from 50 patients diagnosed with non-small-cell lung cancer (NSCLC) and ten adjacent normal lung tissue samples.

On-tissue digestion was performed on a section of the TMA to generate diagnostic peptides from the formalin-crosslinked proteins. After digestion, matrix was applied and spectra were acquired from all biopsy samples in one MS run. A serial section of the TMA was acquired for H&E staining and histological examination by a pathologist. Spectra were classified according to the histological area they were obtained from, and statistical analyses were carried out to develop classification algorithms to predict which biopsy correlated to which diagnosis (specifically adenocarcinoma or squamous

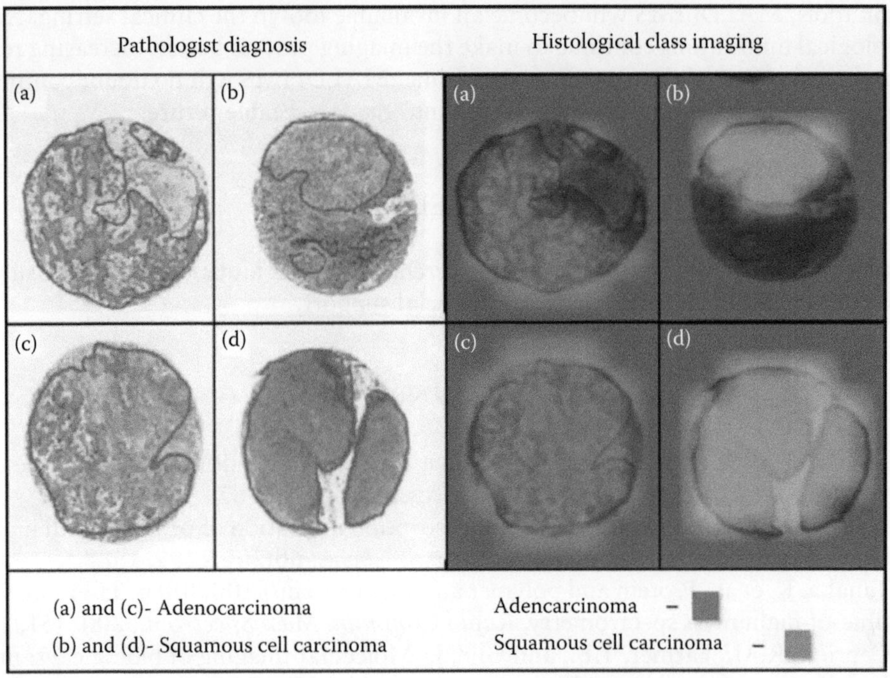

FIGURE 4.9 On-tissue digestion of lung tumor TMA allows for high-throughput diagnostic analyses. Shown is a visual representation of the statistical classification of four biopsies compared to the marking and diagnosis based on histology. (Reproduced from Groseclose, M.R. et al., *Proteomics*, 8, 3715, 2008. With permission.)

cell carcinoma). The analysis is illustrated in Figure 4.9. As shown, the statistical classification made by the support vector machine (SVM) of four biopsies is in agreement with the diagnosis made by the pathologist. This model classified adenocarcinoma spectra with an accuracy of 97.9% and squamous cell carcinoma with an accuracy of 98.6%. This approach combines histology-directed protein profiling with a visualization of the statistical results on hundreds of samples at once. While much work still needs to be done in terms of validation and analysis of fresh frozen samples, the potential of using this technology as a fast, high-throughput, objective tool for the rapid diagnosis of lung tumor biopsies is enormous.

4.8 CONCLUSIONS

The use of MALDI IMS for protein analysis and biomarker discovery has been described. In the relatively short time that this technology has been around, it has enjoyed rapid and widespread use. Unparalleled as a discovery tool, MALDI IMS can simultaneously show the location and state of hundreds of proteins in a given tissue section at one time. This capability has been demonstrated in applications involving understanding normal human development and pathological conditions from Parkinson's disease to cancer. Used in conjunction with histological analysis, protein signatures can be correlated with tissue structures and probed using various statistical algorithms to generate lists for biomarker discovery. As biomarkers already discovered are validated and developed into clinically relevant tools, MALDI IMS will become an invaluable tool in the clinical setting. And as technological innovations continue to make the imaging process better, increasing resolution and sensitivity and decreasing analysis time, MALDI IMS will no doubt continue to be applied to new biological challenges well into the foreseeable future.

ACKNOWLEDGMENT

The authors would like to thank Nipun Merchant for the human pancreas tissue and NIH/NIGMS grant R01 GM58008 for financial support.

REFERENCES

1. Karas, M. et al., Matrix-assisted ultraviolet laser desorption of non-volatile compounds, *Int. J. Mass Spectrom. Ion Processes*, 78, 53, 1987.
2. Karas, M. and Hillenkamp, F., Laser desorption ionization of proteins with molecular masses exceeding 10,000 daltons, *Anal. Chem.*, 60(20), 2299, 1988.
3. Tanaka, K. et al., Protein and polymer analyses up to m/z 100,000 by laser ionization time-of-flight mass spectrometry, *Rapid Commun. Mass Spectrom.*, 2(8), 151, 1988.
4. Caprioli, R.M., Farmer, T.B., and Gile, J., Molecular imaging of biological samples: Localization of peptides and proteins using MALDI-TOF MS, *Anal. Chem.*, 69(23), 4751, 1997.
5. Chaurand, P., Stoeckli, M., and Caprioli, R.M., Direct profiling of proteins in biological tissue sections by MALDI mass spectrometry, *Anal. Chem.*, 71(23), 5263, 1999.
6. Stoeckli, M. et al., Imaging mass spectrometry: A new technology for the analysis of protein expression in mammalian tissues, *Nat. Med.*, 7(4), 493, 2001.
7. Yanagisawa, K. et al., Proteomic patterns of tumour subsets in non-small-cell lung cancer, *Lancet*, 362, 433, 2003.
8. Schwartz, S.A. et al., Protein profiling in brain tumors using mass spectrometry: Feasibility of a new technique for the analysis of protein expression, *Clin. Cancer Res.*, 10, 981, 2004.
9. Schwartz, S.A. et al., Proteomic-based prognosis of brain tumor patients using direct-tissue matrix-assisted laser desorption ionization mass spectrometry, *Cancer Res.*, 65(17), 7674, 2005.
10. Stoeckli, M., Farmer, T.B., and Caprioli, R.M., Automated mass spectrometry imaging with a matrix-assisted laser desorption ionization time-of-flight instrument, *J. Am. Soc. Mass Spectrom.*, 10, 67, 1999.
11. Zhao, J.Y. et al., Application of MALDI-MS/MS molecular imaging software for small molecule profiling in tissue samples, *Proceedings of the 51st ASMS Conference*, Montreal, QC, Canada, 2003.
12. Stoeckli, M. et al., Molecular imaging of amyloid beta peptides in mouse brain sections using mass spectrometry, *Anal. Biochem.*, 311, 33, 2002.
13. Clerens, S., Ceuppens, R., and Arckens, L., *CreateTarget* and *Analyze This!*: New software assisting imaging mass spectrometry on Bruker Reflex IV and Ultraflex II instruments, *Rapid Commun. Mass Spectrom.*, 20(20), 3061, 2006.
14. Pedrioli, P.G.A. et al., A common open representation of mass spectrometry data and its application to proteomics research, *Nat. Biotech.*, 22(11), 1459, 2004.

15. Jardin-Mathe, O. et al., MITICS (MALDI Imaging Team Imaging Computing System): A new open source mass spectrometry imaging software, *J. Proteomics*, 71, 332, 2008.
16. Nye, G.J., Norris, J.L., and Nickerson, S., Optimization of the performance of automated matrix spraying for MALDI imaging mass spectrometry, *Proceedings of the. 54th ASMS Conference*, Seattle, WA, 2006.
17. Baluya, D.L., Garrett, T.J., and Yost, R.A., Automated MALDI matrix deposition method with inkjet printing for imaging mass spectrometry, *Anal. Chem.*, 79(17), 6862, 2007.
18. Dai, Y., Whittal, R.M., and Li, L., Confocal fluorescence microscopic imaging for investigating the analyte distribution in MALDI matrices, *Anal. Chem.*, 68(15), 2494, 1996.
19. Garden, R.W. and Sweedler, J.V., Heterogeneity within MALDI samples as revealed by mass spectrometric imaging, *Anal. Chem.*, 72(1), 30, 2000.
20. Bouschen, W. and Spengler, B., Artifacts of MALDI sample preparation investigated by high-resolution scanning microprobe matrix-assisted laser desorption/ionization (SMALDI) imaging mass spectrometry, *Int. J. Mass Spectrom.*, 266(1–3), 129, 2007.
21. Hankin, J.A., Barkley, R.M., and Murphy, R.C., Sublimation as a method of matrix application for mass spectrometric imaging, *J. Am. Soc. Mass Spectrom.*, 18(9), 1646, 2007.
22. Puolitaival, S.M. et al., Solvent-free matrix dry-coating for MALDI imaging of phospholipids, *J. Am. Soc. Mass Spectrom.*, 19(6), 882, 2008.
23. Aerni, H.-R., Cornett, D.S., and Caprioli, R.M., Automated acoustic matrix deposition for MALDI sample preparation, *Anal. Chem.*, 78(3), 827, 2006.
24. Groseclose, M.R. et al., High-throughput proteomic analysis of formalin-fixed paraffin-embedded tissue microarrays using MALDI imaging mass spectrometry, *Proteomics*, 8(18), 3715, 2008.
25. Agar, N.Y.R. et al., Matrix solution fixation: Histology-compatible tissue preparation for MALDI mass spectrometry imaging, *Anal. Chem.*, 79(19), 7416, 2007.
26. Seeley, E.H. et al., Enhancement of protein sensitivity for MALDI imaging mass spectrometry after chemical treatment of tissue sections, *J. Am. Soc. Mass Spectrom.*, 19(8), 1069, 2008.
27. Lemaire, R. et al., MALDI-MS direct tissue analysis of proteins: Improving signal sensitivity using organic treatments, *Anal. Chem.*, 78(20), 7145, 2006.
28. Chaurand, P. et al., Integrating histology and imaging mass spectrometry, *Anal. Chem.*, 76, 1145, 2004.
29. Schwartz, S.A., Reyzer, M.L., and Caprioli, R.M., Direct tissue analysis using matrix-assisted laser desorption/ionization mass spectrometry: Practical aspects of sample preparation, *J. Mass Spectrom.*, 38, 699, 2003.
30. Xu, B.J. et al., Direct analysis of laser capture microdissected cells by MALDI mass spectrometry, *J. Am. Soc. Mass Spectrom.*, 13(11), 1292, 2002.
31. Laurent, C. et al., Direct profiling of the cerebellum by matrix-assisted laser desorption/ionization time-of-flight mass spectrometry: A methodological study in postnatal and adult mouse, *J. Neurosci. Res.*, 81, 613, 2005.
32. Crecelius, A.C. et al., Three-dimensional visualization of protein expression in mouse brain structures using imaging mass spectrometry, *J. Am. Soc. Mass Spectrom.*, 16, 1093, 2005.

33. Meistermann, H. et al., Biomarker discovery by imaging mass spectrometry: Transthyretin is a biomarker for gentamicin-induced nephrotoxicity in rat, *Mol. Cell. Proteomics*, 5, 1876, 2006.
34. Corbin, B.D. et al., Metal chelation and inhibition of bacterial growth in tissue abscesses, *Science*, 319, 962, 2008.
35. Thibault, D.B. et al., MALDI tissue profiling of integral membrane proteins from ocular tissues, *J. Am. Soc. Mass Spectrom.*, 19(6), 814, 2008.
36. Altelaar, A.F.M. et al., High-resolution MALDI imaging mass spectrometry allows localization of peptide distributions at cellular length scales in pituitary tissue sections, *Int. J. Mass Spectrom.*, 260(2–3), 203, 2007.
37. Chaurand, P. et al., Monitoring mouse prostate development by profiling and imaging mass spectrometry, *Mol. Cell. Proteomics*, 7(2), 411, 2008.
38. Cornett, D.S. et al., A novel histology directed strategy for MALDI-MS tissue profiling that improves throughput and cellular specificity in human breast cancer, *Mol. Cell. Proteomics*, 5(10), 1975, 2006.
39. Stoeckli, M. et al., Imaging of a [beta]-peptide distribution in whole-body mice sections by MALDI mass spectrometry, *J. Am. Soc. Mass Spectrom.*, 18(11), 1921, 2007.
40. Rohner, T.C., Staab, D., and Stoeckli, M., MALDI mass spectrometric imaging of biological tissue sections, *Mech. Ageing Dev.*, 126(1), 177, 2005.
41. Khatib-Shahidi, S. et al., Direct molecular analysis of whole-body animal tissue sections by imaging MALDI mass spectrometry, *Anal. Chem.*, 78(18), 6448, 2006.
42. Yang, E.C.C. et al., Protein expression profiling of endometrial malignancies reveals a new tumor marker: Chaperonin 10, *J. Proteome Res.*, 3, 636, 2004.
43. Taguchi, F. et al., Mass spectrometry to classify non-small-cell lung cancer patients for clinical outcome after treatment with epidermal growth factor receptor tyrosine kinase inhibitors: A multicohort cross-institutional study, *J. Natl. Cancer Inst.*, 99(11), 838, 2007.
44. Sinha, T.K. et al., Integrating spatially resolved three-dimensional MALDI IMS with in vivo magnetic resonance imaging, *Nat. Methods*, 5(1), 57, 2008.
45. Reyzer, M.L. et al., Early changes in protein expression detected by mass spectrometry predict tumor response to molecular therapeutics, *Cancer Res.*, 64, 9093, 2004.
46. Lemaire, R. et al., Specific MALDI imaging and profiling for biomarker hunting and validation: Fragment of the 11S proteasome activator complex, Reg alpha fragment, is a new potential ovary cancer biomarker, *J. Proteome Res.*, 6, 4127, 2007.
47. Amann, J.M. et al., Selective profiling of proteins in lung cancer cells from fine-needle aspirates by matrix-assisted laser desorption ionization time-of-flight mass spectrometry, *Clin. Cancer Res.*, 12(17), 5142, 2006.
48. Skold, K. et al., Decreased striatal levels of PEP-19 following MPTP lesion in the mouse, *J. Proteome Res.*, 5, 262, 2006.
49. Nilsson, A. et al., Increased striatal mRNA and protein levels of the immunophilin FKBP-12 in experimental Parkinson's disease and identification of FKBP-12-binding proteins, *J. Proteome Res.*, 6, 3952, 2007.
50. Burnum, K.E. et al., Imaging mass spectrometry reveals unique protein profiles during embryo implantation, *Endocrinology*, 149(7), 3274, 2008.
51. Andersson, M. et al., Imaging mass spectrometry of proteins and peptides: 3D volume reconstruction, *Nat. Methods*, 5(1), 101, 2008.
52. Schetz, M. et al., Drug-induced acute kidney injury, *Curr. Opin. Crit. Care*, 11, 555, 2005.

53. Charlwood, J. et al., Proteomic analysis of rat kidney cortex following treatment with gentamicin, *J. Proteome Res.*, 1, 73, 2002.
54. Bandara, L.R. et al., A correlation between a proteomic evaluation and conventional measurements in the assessment of renal proximal tubular toxicity, *Toxicol. Sci.*, 73, 195, 2003.
55. Thiery, G. et al., Multiplex target protein imaging in tissue sections by mass spectrometry--TAMSIM, *Rapid Commun. Mass Spectrom.*, 21(6), 823, 2007.
56. Yang, J. et al., Towards LDI imaging mass spectrometry via molecular interaction and photo labile tags, *Proceedings of the 55th ASMS Conference*, Indianapolis, IN, 2007.
57. Groseclose, M.R. et al., Identification of proteins directly from tissue: *In situ* tryptic digestions coupled with imaging mass spectrometry, *J. Mass Spectrom.*, 42(2), 254, 2007.
58. Xu, H. et al., Antigen retrieval for proteomic characterization of formalin-fixed and paraffin-embedded tissues, *J. Proteome Res.*, 7, 1098, 2008.
59. Shi, S.-R. et al., Protein extraction from formalin-fixed, paraffin-embedded tissue sections: Quality evaluation by mass spectrometry, *J. Histochem. Cytochem.*, 54(6), 739, 2006.
60. Ronci, M. et al., Protein unlocking procedures of formalin-fixed paraffin-embedded tissues: Application to MALDI-TOF imaging MS investigations, *Proteomics*, 8, 3702, 2008.

CHAPTER 5

Secondary-Ion Mass Spectrometry

Leo M.L. Nollet

CONTENTS

5.1 SECONDARY-ION MASS SPECTROMETRY

Secondary-ion mass spectrometry (SIMS) is the most sensitive surface analysis technique providing quantifiable data to the ppb level and surface specificity to the uppermost monolayer. It can be used to analyse most solid vacuum compatible materials including metals, semiconductors, ceramics, polymers, and biological material.

In secondary-ion mass spectrometry the composition of solid surfaces and thin films is analysed by sputtering the surface of the sample with a focused primary ion beam and collecting and analysing ejected secondary ions. The system is maintained under ultra-high vacuum to prevent atmospheric interference with the primary and secondary ions.

The mass/charge ratios of these secondary ions are measured with a mass spectrometer to determine the elemental, isotopic, or molecular composition of the surface to a depth of 1 to 2 nm. Due to the large variation in ionisation probabilities among different materials, SIMS is generally considered to be a qualitative technique, although quantitation is possible with the use of standards. SIMS is a very sensitive surface analysis technique, with elemental detection limits ranging from parts per million to parts per billion.

The interaction of the primary ion beam with the sample (under vacuum) provides sufficient energy to ionise many elements. If the primary beam is composed of positively charged ions, the resultant ionisation favours the production of negative ions; primary beams of negative ions favour the generation of positive ions. Although most atoms and molecules removed from the sample by the interaction of the primary beam and the sample surface are neutral, a percentage of these are ionised. These ions are then accelerated, focused, and analysed by a mass spectrometer.

When a primary ion impacts the specimen it gives up its energy through a series of collisions with target atoms which recoil and, in turn, collide with their neighbours. If they have sufficient energy, these will also go on to collide with other atoms, resulting in a collision cascade. At some point, the cascade may intersect the surface, and the energetic particle, instead of colliding with a nearby atom, may escape from the material.

If conditions are favourable the emitted particle may be in an ionised state. Generally, although the primary ion may penetrate many nanometres into the target, the ejected ions come from only the uppermost few atomic layers.

There are two classes of SIMS instruments: static SIMS (time-of-flight secondary-ion mass spectrometry: TOF-SIMS), which uses a pulsed primary ion beam with limited dose ($<10^{12}$ ions cm^{-2}) to reveal the molecular composition of the top surface, and dynamic SIMS (nano-SIMS), using an unlimited primary-ion dose for in-depth elemental distribution analysis. The main advantage of the latest generation of nano-SIMS spectrometers (e.g. Cameca, Nano-SIMS 50L) over TOF-SIMS is very high mass resolution.

Time-of-flight secondary-ion mass spectrometry (TOF-SIMS) is a surface-sensitive analytical method that uses a pulsed ion beam to remove molecules from the surface of the sample. The particles are removed from atomic monolayers on the surface (secondary ions). These particles are then accelerated into a "flight tube" and their mass is determined by measuring the exact time at which they reach the detector (i.e. time-of-flight). Three operational modes are available using TOF-SIMS: surface spectroscopy, surface imaging, and depth profiling.

Nanoscale secondary-ion mass spectrometry (nano-SIMS or nano secondary-ion mass spectrometry) is a nanoscopic scale resolution chemical imaging mass spectrometer based on secondary-ion mass spectrometry (Figure 5.1).

Nano-SIMS not only refers to the technique used, but also the mass spectrometer specialized for this method.

Depending on the SIMS type, there are three basic analysers available: sector, quadrupole, and time-of-flight. A sector field mass spectrometer uses a combination of an electrostatic analyser and a magnetic analyser to separate the secondary ions by their

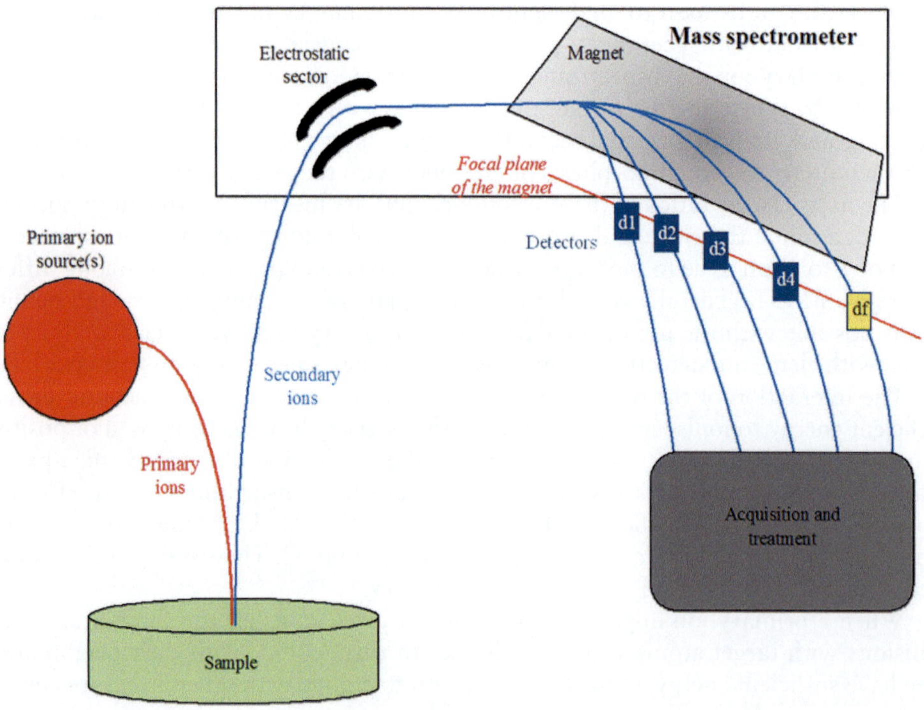

FIGURE 5.1 Scheme of nano-SIMS.

mass-to-charge ratio. A quadrupole mass analyser separates the masses by resonant electric fields, which allow only the selected masses to pass through. The time-of-flight mass analyser separates the ions in a field-free drift path according to their velocity. Since all ions possess the same kinetic energy the velocity and therefore time of flight vary according to mass. It requires pulsed secondary-ion generation using either a pulsed primary-ion gun or a pulsed secondary-ion extraction. It is the only analyser type able to detect all generated secondary ions simultaneously, and is the standard analyser for static SIMS instruments.

Liquid secondary-ion mass spectrometry (LSIMS) is an ionization technique sometimes called dynamic FAB. The nonvolatile sample is dissolved in a liquid matrix (usually glycerol) before introduction into the source.

5.2 APPLICATIONS

5.2.1 Nano-SIMS

Cereals are an important source of selenium (Se) to humans; in contrast, arsenic (As) is toxic and may accumulate in rice grain at levels that may pose a health risk. Knowledge of the localisation of selenium and arsenic within the cereal grain will aid understanding of their deposition patterns and the impact of processes such as milling.

Nano-SIMS was used to determine the localisation of Se in wheat (*Triticum aestivum*) and As in rice (*Oryza sativa*) (1). Combined synchrotron X-ray fluorescence (S-XRF) and nano-SIMS analysis utilized the strengths of both techniques.

Selenium was concentrated in the protein surrounding the starch granules in the starchy endosperm cells and more homogeneously distributed in the aleurone cells but with Se-rich hotspots. Arsenic was concentrated in the subaleurone endosperm cells in association with the protein matrix rather than in the aleurone cells. Nano-SIMS indicated that the high intensity of As identified in the S-XRF image was localised in micron-sized hotspots near the ovular vascular trace and nucellar projection.

There was good quantitative agreement between nano-SIMS and S-XRF.

Insufficient iron (Fe) is one of the most prevalent micronutrient deficiencies in humans, with billions of people affected (2). Cereal grains are an important source of Fe for humans, but the bioavailability of Fe in cereals is generally low. Information regarding the cellular and sub-cellular localisation of Fe in wheat grain will aid in optimising nutrient delivery for human health. Nano-SIMS was used to map the distribution of Fe in the aleurone layer and in the endosperm of immature wheat grain. Iron was shown to be localised strongly in the phytin globoids in the aleurone cells and to a lesser extent in the cytoplasm around the starch granules in the endosperm.

5.2.2 TOF-SIMS

Ahn et al. (3) developed a simple and multiplexed assay of aflatoxins using time-of-flight secondary-ion mass spectrometry. By simply adsorbing either a single analyte or mixed ones onto a gold substrate, the corresponding secondary molecular ions ([M+H]+) were clearly observed in a single mass spectrum. As a result of concentration-dependent peak intensity, quantitative and multiplexed analysis of different aflatoxin analogues from corns was accomplished with an immunoaffinity column and TOF-SIMS analysis, which showed a good correlation with HPLC data. The detection sensitivity was estimated to be as low as 10 ng mL^{-1}.

Proof-of-principle experiments were worked out utilizing commercial cellulase and laccase enzymes, which are known to modify major polymeric components of wood (i.e. cellulose and lignin, respectively) (4). Enzyme activity was assessed through principle component analysis (PCA) as well as through peak ratios intended to measure selective enzymatic wood degradation. Spectral reproducibility of the complex wood substrates was found to be within 5% relative standard deviation (RSD), allowing for relative quantification of changes in wood composition. Procedures were also presented to identify and avoid the influence of mass interferences from protein adsorption by the enzyme solutions. The activity of a cellulase cocktail is clearly evident through the TOF-SIMS spectra and is supported by high-pressure liquid chromatography (HPLC) measurements of sugar release and by complementary X-ray photoelectron spectroscopy (XPS) measurements of the wood surfaces. Laccase activity, which is mediated through small organic molecules, can be detected in the TOF-SIMS spectra through a decrease in G and S lignin peaks.

Pollen collected by honey bees contains a substantial amount of nutrients and has a high nutritive value. However, it can be difficult to digest and absorb a high level of nutrients due to the complex wall of bee pollen.

It was observed that amino acids were mostly distributed inside the cell wall of lotus bee pollen, rape bee pollen, apricot bee pollen, wuweizi bee pollen, and camellia bee pollen, using time-of-flight secondary-ion mass spectrometry (5). Five species of bee pollen were wall disrupted with a combination of ultrasonication and high shear technique (US-HS). After this treatment, bee pollen walls were completely broken into fragments, and a large number of nutrients were released. The amino acid, fatty acid, protein, crude fat, reducing sugar, β-carotene, calcium, iron, zinc, and selenium content increased after wall disruption.

Time-of-flight secondary-ion mass spectrometry seemed to be a reliable mapping technique for determining the distribution of food ingredients.

Fundamental advances in secondary-ion mass spectrometry allow for the examination and characterisation of lipids directly from biological materials (6). The successful application of SIMS-based imaging in the investigation of lipids directly from tissue and cells is demonstrated by Passarelli and Winograd (2011). In their review, they examine the use of cluster ion sources and cryogenically compatible sample handling for improved ion yields and to expand the application potential of SIMS. Methodological improvements, including pre-treating the sample to improve ion yields and protocol development for three-dimensional analyses (i.e. molecular depth profiling), were also included in their discussion. New high-performance SIMS instruments showcasing the most advanced instrumental developments, including tandem MS capabilities and continuous ion beam compatibility, were described.

Meat is a highly structured product consisting of muscle fibres, lipids, and connective tissue sheaths. These different compounds are suspected of modulating the diffusion of digestive enzymes in meat during its gastric digestion. To follow the pepsin diffusion in digested meat and better understand the role of meat structure organization, time-of-flight secondary-ion mass spectrometry imaging was used to detect pepsin in food (7). Two beef muscle samples (1.5 × 1.5 × 2.5 cm) were incubated in a pepsin solution at pH 2. After two hours' incubation at 37°C under agitation, the samples were cryofixed; histological sections were collected on glass slides and used for chemical imaging. The entire surface area of the sections (1.5 × 1.5 cm) was mapped by TOF-SIMS imaging. The use of TOF-SIMS allowed the visualization of pepsin in a food.

There are several ways to improve the quality of this type of meat, including changing the concentrations of individual molecular components. Such important components of meat are inter alia, cholesterol, vitamin E, and some fatty acids such as ω-3 and ω-6. Manipulation of ingredient levels may be achieved by enriching chicken feed with elements of different types such as vegetable oils, garlic, or selenium. Various biochemical and biophysical methods have been used to study the quality of different meat types, especially broiler meat.

The use of high-resolution time-of-flight secondary-ion mass spectrometry was demonstrated to assess how variations in animal nutrition affect concentrations of specific lipids in the meat, such as cholesterol and vitamin E (8). Four different dietary treatments were tested. Feed for animals in the first group was supplemented with soy oil in 50%, the second group's feed was supplemented with linseed oil in 50%, a combination of these two oils in the proportion of 44%:56% was used for the third group, and in the reference group, animals were fed with beef tallow. From each group, four individuals were selected for further analysis. Positive and negative ion mass spectra were generated from the pectoralis superficialis muscle tissue of the left carcass side of each animal. Using TOF-SIMS with a bismuth cluster ion source (Bi3+), and based on characteristic peaks for cholesterol in the positive mode and vitamin E in the negative mode, the relationships of these lipid levels to the various feeding regimens were demonstrated.

Peppers (*Capsicum* spp.) are well-known for their ability to cause an intense burning sensation when eaten. This organoleptic response is triggered by capsaicin and its analogues, collectively called capsaicinoids. In addition to the global popularity of peppers as a spice, there is a growing interest in the use of capsaicinoids to treat a variety of human ailments, including arthritis, chronic pain, digestive problems, and cancer. The cellular localization of capsaicinoid biosynthesis and accumulation has previously been studied by fluorescence microscopy and electron microscopy, both of which require immunostaining. TOF-SIMS has been used to image the distribution of capsaicinoids in the interlocular septum and placenta of *Capsicum chinense* (Scotch bonnet peppers) (9). A unique cryo-TOF-SIMS instrument has been used to prepare and analyse the samples with minimal sample preparation. Samples were frozen in liquid propane, cryosectioned in vacuum, and analysed without exposure to ambient pressure. TOF-SIMS imaging was performed at –110°C using a Bi3+ primary ion beam. Molecular ions for capsaicin and four other capsaicinoids were identified in both the positive and negative TOF-SIMS spectra. The capsaicinoids were observed to be concentrated in pockets between the outer walls of the palisade cells and the cuticle of the septum, as well as in the intercellular spaces in both the placenta and interlocular septum.

An imaging secondary-ion mass spectrometry system has been developed allowing the distribution of elements or ions to be superimposed on an image of the plant cell or tissue generated by ion-induced secondary electrons (10). This system has been evaluated by analysing the aleurone and sub-aleurone cells of mature wheat grain, showing high spatial resolution (100–200 nm) images of O–, PO2–, Mg+, Ca+, Na+, and K+ within the phytate granules of the aleurone, with CN– being diagnostic for proteins and C2– being diagnostic for starch in the starchy endosperm cells.

Time-of-flight secondary-ion mass spectrometry imaging has been used to map flavonoids in fresh seed sections of peas (*Pisum sativum*) and *Arabidopsis thaliana* (11). While for peas a very simple preparation method derived from mammalian tissue imaging could be utilized, several preparation methods had to be tested for the *A. thaliana* seeds before obtaining tissue sections on which the diagnostic ions were not delocalised. For such small and stiff biological material, none of the methods currently used in histology or scanning electron microscopy could be transferred to mass spectrometry imaging. Only the embedding of the fresh seeds in a polyester resin, followed by the analysis of the block after having obtained a flat surface section with a diamond blade, gave sensitive and reproducible results. Several flavonoid ions have been detected in the sections, showing increased concentrations of flavonoids in the seed coats.

5.2.3 LSIMS

Several fractions of oligomeric and polymeric procyanidins have been obtained from grape seeds by gel chromatography on Fractogel TSK (12). Their molecular weights (Mw) were directly determined by liquid secondary-ion mass spectrometry (LSIMS). The authors observed that tannins were molecules with a wide range of Mw: from 290 to 3100 (decamers). Catechins (Mw 290), procyanidin dimers (Mw 578), (-)-epicatechin gallate (Mw 442), procyanidin dimer gallates (Mw 730), procyanidin dimer digallates (Mw 882), procyanidin trimers (Mw 866), and procyanidin trimer gallates (Mw 1018) were identified as $(M-H)^-$ quasimolecular ions by negative LSIMS. Some structures were elucidated by partial thiolysis and enzymatic hydrolysis.

REFERENCES

1. K. L. Moore, M. Schröder, E. Lombi, F.-J. Zhao, S. P. McGrath, M. J. Hawkesford, P. R. Shewry, C. R. M. Grovenor. NanoSIMS analysis of arsenic and selenium in cereal grain. *New Phytologist*, 185 (2), 434–445, 2010.
2. K. L. Moore, F.-J. Zhao, C. S. Gritsch, M. J. Hawkesford, S. P. McGrath, P. R. Shewry, C. R. M. Grovenor. Localisation of iron in wheat grain using high resolution secondary ion mass spectrometry. *Journal of Cereal Science*, 55 (2), 183–187, 2012.
3. J.-H. Ahn, Y.-S. Jeong, T. G. Lee, Y.-P. Kim, H.-S. Kim. Sensitive and multiplexed analysis of aflatoxins using time-of-flight secondary ion mass spectrometry. *BioChip Journal*, 6 (1), 34–40, 2012.
4. R. E. Goacher, E. A. Edwards, A. F. Yakunin, C. A. Mims, E. R. Master. Application of time-of-flight-secondary ion mass spectrometry for the detection of enzyme activity on solid wood substrates. Analytical Chemistry, 84 (10), 4443–4451, 2012.
5. W. Wu, K. Wang, J. Qiao, J. Dong, Z. Li, H. Zhang. Improving nutrient release of wall-disrupted bee pollen with a combination of ultrasonication and high shear technique. *Journal of the Science of Food and Agriculture*, 99 (2), 564–575, 2018.
6. M. K. Passarelli, N. Winograd. Lipid imaging with time-of-flight secondary ion mass spectrometry (ToF-SIMS). *Biochimica et Biophysica Acta (BBA) - Molecular and Cell Biology of Lipids*, 1811 (11), 976–990, 2011.
7. T. Astruc, A. Venien, O. Loison, D. Pavon. Pepsin localization in digested meat by ToF-SIMS analysis. *International Symposium on Food Nutrition and Health*, May 2017, Dalian, China, 2017.
8. M. E. Marzec. Study of cholesterol and vitamin E levels in broiler meat from different feeding regimens by TOF-SIMS. *Biointerphases*, 11, 02A326, 2016. doi: 10.1116/1.4943619
9. B. J. Tyler. ToF-SIMS imaging of capsaicinoids in Scotch Bonnet peppers (*Capsicum chinense*). *Biointerphases*, 11, 02A327, 2016.

10. P. J. Heard, K. A. Feeney, G. C. Allen, P. R. Shewry. Determination of the elemental composition of mature wheat grain using a modified secondary ion mass spectrometer (SIMS). *The Plant Journal*, 30 (2), 237–245, 2002.
11. A. Seyer, J. Einhron, A. Brunelle, O. Laprévote. Localization of flavonoids in seeds by cluster time-of-flight secondary ion mass spectrometry imaging. Analytical Chemistry, 256 (2), 145–152, 2014.
12. V. A. P. deFreitas, Y. Glories, G. Bourgeois, C. Vitry. Characterisation of oligomeric and polymeric procyanidins from grape seeds by liquid secondary ion mass spectrometry. *Phytochemistry*, 49 (5), 1435–1441, 1998.

CHAPTER 6

Desorption Electrospray Ionization Mass Spectrometry Imaging in Food Applications

Semih Ötleş and Vasfiye Hazal Özyurt

CONTENTS

6.1 INTRODUCTION

Desorption electrospray ionization (DESI) is a powerful and versatile ambient ionization surface analysis technique. In this ionization, molecules are firstly desorbed and then ionized from the surface under ambient conditions. After that, the ionized molecules are analyzed by mass spectrometry (Claude, Jones, & Pringle, 2017). The DESI technique consists of charged solvent droplets and then these ionized droplets are transported toward the atmospheric pressure inlet of the mass spectrometer (Takáts, Wiseman, & Cooks, 2005). Among mass ionization sources, DESI has both advantages and disadvantages according to variety of molecules (Garza et al., 2018; Ifa et al., 2011).

Mass spectrometry imaging (MSI) has gained popularity in recent years (Boughton, Thinagaran, Sarabia, Bacic, & Roessner, 2016) since it enables label-free detection and mapping of a wide range of molecules at complex surfaces (Dong et al., 2016). DESI-MSI has been applied in several fields such as pharmaceuticals (Chen, Talaty, Takats, & Cooks, 2005), environmental analysis (Campbell, Ton, & Mulligan, 2011), food safety testing (Garcia-Reyes, Jackson, Molina-diaz, & Cooks, 2009), forensics (Takáts et al., 2005), and biological analyses (Eberlin, Ferreira, Dill, Ifa, & Cooks, 2011).

Here, we describe the workflow of DESI-MSI and how to obtain DESI-MSI image maps of food.

6.2 MASS SPECTROMETRY IMAGING

Mass spectrometry imaging (MSI) is used to elucidate chemical distributions of samples and to provide spatial context (Garza et al., 2018; Veličković & Anderton, 2017). MSI is a label-free analysis and allows the simultaneous multiplex measurement of hundreds to possibly thousands of analytes in a single imaging experiment (Boughton et al., 2016).

An MSI workflow is clear. There are four steps for basic MSI:

(1) Sample selection and preparation
(2) Desorption and ionization
(3) Mass analysis
(4) Image registration and data analysis

Firstly, samples are commonly sectioned (e.g., frozen tissue) or excised (e.g., agar supported material). Secondly, they are added on a substrate. Then these samples are probed by an ionization source. After this, a series of mass spectra is generated from predefined areas (Dong et al., 2016). There are two different types of modes: microprobe mode and microscope mode. A microprobe mode is the most common mode. In this mode, sequential individual spectra are collected and combined into a single dataset. The microscope mode is an experimental mode (Boughton et al., 2016).

There are numerous combinations of ionization sources and mass analyzers and each has its own advantages and limitations. Essentially, there are the different types of desorption/ionization methods utilized in MSI (Boughton et al., 2016). Matrix-assisted laser desorption/ionization (MALDI) is the most widely used MS imaging technique in vacuum (Wu, Dill, Eberlin, Cooks, & Ifa, 2012), while one of most widely used ambient ionization MS techniques is desorption electrospray ionization (DESI) MS imaging (Boughton et al., 2016; Ifa & Eberlin, 2016). Alternative matrix-free approaches are using laser desorption ionization (LDI), secondary-ion mass spectrometry (SIMS), and coupled sources including laser ablation electrospray ionization (LAESI) are other types of ion sources (Boughton et al., 2016). Moreover, many other MSI sources are still emerging, and they are usually derivatives or modifications of the above-mentioned approaches. Among them it is worth mentioning laser ablation inductively coupled plasma (LA-ICP), liquid extraction surface analysis (LESA), direct analysis in real time (DART), and nano-desorption electrospray ionization (Nano-DESI) (Dong et al., 2016; Sica, Raja, El-Elimat, & Oberlies, 2014; Thunig, Hansen, & Janfelt, 2011).

6.3 DESORPTION ELECTROSPRAY IONIZATION-MASS SPECTROSCOPY IMAGING

DESI is an ambient ionization technique used for the mass spectrometric imaging of samples (Garza et al., 2018; Lewis et al., n.d.; Takáts, Wiseman, Gologan, & Graham Cooks, 2004). There are different types of ion sources available. DESI ionizes compounds with little sample preparation and without a vacuum system, and it involves spraying electrically charged solvent droplets onto a sample surface, accompanied by a high-velocity gas jet. The generally accepted model for DESI analyte ionization involves analyte dissolving in a thin layer of the DESI solvent before being ejected in secondary droplets by impact from the primary spray. The charged secondary droplets evolve into gas-phase ions similar to the theorized electrospray mechanism due to electrostatic and pneumatic forces. The desorbed gas-phase ions are then sucked into a mass spectrometer for detection through a custom-built electrically charged metal inlet. With a soft-ionization mechanism, DESI can be used to see molecular ions of larger biological molecules as well as multiply charged ions sampled directly from intact biological tissue. DESI could ionize both polar and nonpolar molecules such as alkaloids, peptides, and proteins that are present on varied surfaces including metals, polymers, and minerals. It could be applied for both quantification and qualification of molecules (Araújo et al., 2017; Garza et al., 2018).

DESI relies on a soft ionization technique, which delivers mass spectra with very low fragmentation in either positive or negative ionization mode. The DESI source can be connected to existing mass spectrometers. It is called DESI-MSI. DESI-MSI utilizes the original principle of electrospray ionization, but in this case solvent droplets are desorbed directly on the sample surface. It requires basic sample preparation and allows repeated measurements on samples. Then it results in a simplified analytical procedure for a rapid spatial and temporal identification of chemicals in samples all under ambient conditions (Griffiths, Kocurek, & Cooper, 2018). The workflow of DESI is that high-velocity ionized solvent droplets desorb the analytes directly from the sample surface. Solvents are electro-sprayed under high voltage through an emitter capillary producing charged "primary" droplets, and then directed toward the sample. Metabolites located on the sample surface are desorbed into gaseous "secondary" droplets delivering molecular ions entering the MS inlet where *m/z* values are measured (Mohana Kumara et al., 2016; Towers, Karancsi, Jones, Pringle, & Claude, 2018).

According to the polarity, different spray solvents such as MeOH or ACN with the addition of 2–5% H_2O can be used for imaging, and different solvent compositions have been employed according to the specific needs and application type (Takáts et al., 2005). However, in some cases, ionization efficiency and its selectivity should be increased by using specific ionization-aiding treatments. A reactive reagent should also be added to the spray solvent to selectively improve the ionization efficiency of analytes (Dong et al., 2016). In addition to solvents, salt adducts to solvents have shown to have a high degree of tolerance for formation and ionization suppression (Takáts et al., 2005, 2004). The correct adjustment of different parameters is going to affect the analysis and the final imaging resolution.

The most critical parameters are the electrospray solvent flow rate and voltage, and the distance of sample surface to the DESI spray nozzle tip and to the MS-inlet, as well as their respective incident and collection degree angles (α and β) (Takáts et al., 2005, 2004).

Sample preparation for DESI-MSI may change depending on sample types. Sample preparation from plant tissues is more challenging than those for mammalian tissues (Dong et al., 2016). There is no need for preparation of samples with smooth and regular surfaces, while cryosections can be prepared in non-flat surfaces as freezing. In plants the sample preparation step is different according to sample type. Since flowers and leaves are soft, irregular, and have a very absorbent surface, signal during imaging is low and instable (Schwartz, Reyzer, & Caprioli, 2003). The thickness of sample is also important for the imaging process. For mammalian tissues, section thickness between 5 and 20 μm is recommended for the analysis of low molecular weight molecules, and <5μm thickness for high molecular weight proteins ($m/z > 9000$) (Dong et al., 2016).

DESI-MSI is carried out by directly scanning the unmodified sample in the *x*- and *y*-directions through an impinging spray of charged droplets; the chemical information obtained can then be plotted as two-dimensional images recording the abundance of specific ions (Eberlin et al., 2011). DESI-MSI contains some information and thus requires intensive analysis for data extraction, visualization, and interpretation. For that reason, advanced software and computational data analysis techniques are needed to extract results from the data. Some software packages are BioMap (Novartis, Basel, Switzerland), Datacube Explorer (FOM-AMOLF, Amsterdam, Netherlands), FlexImaging and

TABLE 6.1 Application DESI-MSI in Food

Sample Type	Spray Solvent	Image Resolution	Analytes	Ionization Mode	References
Red macroalga (*Callaphycus serratus*)	100 μM NH_4Cl in MeOH	200 μm	Bromophycolides	Negative	(Lane et al., 2009)
Edible oils and margarine	Water/methanol 1:1 (v/v)	–	Triglycerides	Positive and negative	(Gerbig & Takats, 2010)
St. John's wort (*H. perforatum*); thorn apple (*Datura stramonium*)	50:50 mixture of MeOH/H_20 with 1% of ammonia 50:50 mixture of MeOH/H20 with 1% of formic acid	100–125 μm	Phloroglucinols, flavonoids, naphthodianthrones, saccharides, alkaloids	Negative Positive	(Thunig et al., 2011)
Barley (*Hordeum vulgare*)	MeOH and water in a 4:1	100–200 μm	Hydroxynitrile glucosides	Negative	(Li, Bjarnholt, Hansen, & Janfelt, 2011)
Katsura tree (*C. japonicum*); American sweetgum (*Liquidambar styraciflua*)	1% formic acid in methanol/water (70:30) 0.1% ammonium acetate in methanol/water (70:30)	130–310 μm	Chlorophyll catabolites	Positive and Negative	(Müller, Oradu, Ifa, Cooks, & Kr, 2011)
Myristica malabarica	Methanol and water (9:1 v/v)	250 μm	Alkaloid	Negative	(Ifa et al., 2011)
Potato (*Solanum tuberosum*); Gingko (*Gingko biloba L.*); Strawberry (*Fragaria 9 ananassa* Duch)	Methanol, acetonitrile, and mixtures of both	150–200 μm	Glycoalkaloids, flavonoids, sugars and anthocyanidin	Negative and positive	(Cabral, Mirabelli, Perez, & Ifa, 2013)

(*Continued3*)

TABLE 6.1 (CONTINUED) Application DESI-MSI in Food

Sample Type	Spray Solvent	Image Resolution	Analytes	Ionization Mode	References
C. roseus	Methanol, methanol–water (3:1 v/v), and acetonitrile	400 µm	Serpentine; vindoline; catharanthine; 19-S-vindoline; like-catharanthine; perivine; alstonine; tabersonine isomers; dihydrotabersonine; ajmalicine; hydroxytarbersonine; decaetoxyvindoline; S-adenosylmethionine; akuamimicine; methoxytabersonine; lochnerinine; echitovenine; deacetoxyvindoline; vindolinine; strictosidine; anhydrovinblastine; vinblastine	Positive	(Hemalatha & Pradeep, 2013)
H. perforatum	$CHCl_3$–ACN–H_2O (1:1:0.04)	50–150µm	Tetracosanic acid; hexacosanoic acid; octacosanoic acid; melissic acid; quercetin; quercitrin; isobaric mixture of isoquercitrin; hyperoside; hypericin; protohypericin; pseudohypericin; protopseudohypericin; C26, C28 and C30 fatty acids; hyperforin; rutin; hyperfirin; adhyperfin	Negative	(Li, Hansen, & Janfelt, 2013)
D. binectariferum	Methanol	250 µm	Rohitukine; acetylated and glycosylated rohitukine	Positive	(Kumara, Srimany, Ravikanth, Shaanker, & Pradeep, 2015)
Grape leaf petiole	ACN:H_2O (4:1)	High resolution	Tartaric acid	Negative	(Dong, Guella, Mattivi, & Franceschi, 2015)
Zebrafish	Methanol	200 µm	Phospholipids	Positive	(Perez, Tata, Campos, Peng, & Ifa, 2017)

ClinProTools (Bruker Daltonik, Bremen, Germany), HDI (high-definition MALDI MS imaging) coupled to MassLynx and MarkerLynx (Waters, Manchester UK), ImageQuest (Thermo Scientific, Waltham, MA, USA), MALDIVision (PREMIER Biosoft), Metabolite Imager (University of Texas), MIRION (Justus Liebig University), MSiReader (North Carolina State University), OpenMSI (Lawrence Berkeley National Lab, CA, USA, http://openmsi.nersc.gov), SCiLS Lab (SCiLS Bremen, Germany), and TissueView (AB Sciex, based on BioMap) (Boughton et al., 2016).

Due to its easy workflow and versatility, DESI-MSI has successfully been applied to many different research fields (Gerbig & Takats, 2010). Herein we discuss the present applications of DESI-MSI in food.

6.4 FOOD APPLICATIONS

Some publications focused on DESI-MSI are summarized in Table 6.1. Table 6.1 covers the application of a new technology to samples, and the development of new techniques and methods to examine metabolites. DESI-MSI helps to identify the fundamental biology, chemical structure, and surface metabolites, and it is applied to measure stress responses. Moreover, this technique provides insights into normal plant growth.

There are a few publications about the application of DESI-MSI in the localization of carbohydrates due to their low selectivity and the sensitivity of this technique in terms of these molecules (Cabral et al., 2013; Thunig et al., 2011). In mammalian tissues, DESI-MSI is the most important application for the determination of lipid localization (Perez et al., 2017). Terpenoids, a large and diverse class of metabolites, can be employed for various functions such as the growth and development of plants. Thunig et al. (2011) investigated the distribution of terpenoids in *Hypericum perforatum*. In plants (*Conium maculatum*, *Datura stramonium*, and *Atropa belladonna*), alkaloids, which act as defense compounds against pathogenic organisms and herbivores, have been investigated (Talaty, Takats, & Cooks, 2005). Cabral et al. (2013) employed DESI-MSI for the localization of flavanols in *Ginkgo* leaves and of anthocyanins in strawberry.

6.5 CONCLUSION

DESI-MSI has successfully been used as a rapid method for the trace analysis of components in food. DESI-MSI research will promote the future study for the development of new techniques and will supply instrumentation for chemical mapping of foods. In recent years, the DESI-MSI technique has developed rapidly in food samples.

REFERENCES

Araújo, F. D. S., Vieira, R. L., Molano, E. P. L., Máximo, H. J., Dalio, R. J. D., Vendramini, P. H., ... Eberlin, M. N. (2017). Desorption electrospray ionization mass spectrometry imaging reveals chemical defense of: Burkholderia seminalis against cacao pathogens. *RSC Advances*, 7(48), 29953–29958. doi:10.1039/c7ra03895j

Boughton, B. A., Thinagaran, D., Sarabia, D., Bacic, A., & Roessner, U. (2016). Mass spectrometry imaging for plant biology: A review. *Phytochemistry Reviews*, *15*(3), 445–488. doi:10.1007/s11101-015-9440-2

Cabral, E. C., Mirabelli, M. F., Perez, C. J., & Ifa, D. R. (2013). Blotting assisted by heating and solvent extraction for DESI-MS imaging. *Journal of the American Society for Mass Spectrometry*, *24*, 956–965. doi:10.1007/s13361-013-0616-y

Campbell, I. S., Ton, A. T., & Mulligan, C. C. (2011). Direct detection of pharmaceuticals and personal care products from aqueous samples with thermally-assisted desorption. *Journal of the American Society for Mass Spectrometry*, *22*, 1285–1293. doi:10.1007/s13361-011-0144-6

Chen, H., Talaty, N. N., Takats, Z., & Cooks, R. G. (2005). Desorption electrospray ionization mass spectrometry for high-throughput analysis of pharmaceutical samples in the ambient environment. *Analytical Chemistry*, *77*(21), 6915–6927.

Claude, E., Jones, E. A., & Pringle, S. D. (2017). DESI mass spectrometry imaging (MSI). In L. M. Cole (Ed.), *Imaging Mass Spectrometry* (pp. 65–76). Sheffield, South Yorkshire, UK: Springer. Retrieved from http://www.springer.com/series/7651

Dong, Y., Guella, G., Mattivi, F., & Franceschi, P. (2015). High production of small organic dicarboxylate dianions. *Journal of the American Society for Mass Spectrometry*, *26*, 386–389. doi:10.1007/s13361-014-1065-y

Dong, Y., Li, B., Malitsky, S., Rogachev, I., Aharoni, A., Kaftan, F., ... Franceschi, P. (2016). Sample preparation for mass spectrometry imaging of plant tissues: A review. *Frontiers in Plant Science*, *7*(February). doi:10.3389/fpls.2016.00060

Eberlin, L. S., Ferreira, C., Dill, A. L., Ifa, D. R., & Cooks, R. G. (2011). Desorption electrospray ionization mass spectrometry for lipid characterization and biological tissue imaging. *Biochimica et Biophysica Acta*, *1811*(11), 946–960. doi:10.1038/mp.2011.182

Garcia-Reyes, J. F., Jackson, A. U., Molina-diaz, A., & Cooks, R. G. (2009). Desorption electrospray ionization mass spectrometry for trace analysis of agrochemicals in food. *Analytical Chemistry*, *81*(2), 820–829.

Garza, K. Y., Feider, C. L., Klein, D. R., Rosenberg, J. A., Brodbelt, J. S., & Eberlin, L. S. (2018). Desorption electrospray ionization mass spectrometry imaging of proteins directly from biological tissue sections. *Analytical Chemistry*, *90*(13), 7785–7789. article-commentary. doi:10.1021/acs.analchem.8b00967

Gerbig, S., & Takats, Z. (2010). Analysis of triglycerides in food items by desorption electrospray ionization mass spectrometry. *Rapid Communications in Mass Spectrometry*, *24*, 2186–2192. doi:10.1002/rcm

Griffiths, R. L., Kocurek, K. I., & Cooper, H. J. (2018). Ambient surface mass spectrometry–ion mobility spectrometry of intact proteins. *Current Opinion in Chemical Biology*, *42*, 67–75. doi:10.1016/j.cbpa.2017.11.002

Hemalatha, R. G., & Pradeep, T. (2013). Understanding the molecular signatures in leaves and flowers by desorption electrospray ionization mass spectrometry (DESI MS) imaging. *Journal of Agricultural and Food Chemistry*, *61*(31), 7477–7487.

Ifa, D. R., & Eberlin, L. S. (2016). Ambient ionization mass spectrometry for cancer diagnosis and surgical margin evaluation. *Clinical Chemistry*, *62*(1), 111–123. doi:10.1373/clinchem.2014.237172

Ifa, D. R., Srimany, A., Eberlin, L. S., Naik, H. R., Bhat, V., Cooks, R. G., & Pradeep, T. (2011). Imprint imaging by desorption electrospray ionization mass spectrometry. *Analytical Methods*, *3*, 1910–1912. doi:10.1039/c1ay05295k

Kumara, P. M., Srimany, A., Ravikanth, G., Shaanker, R. U., & Pradeep, T. (2015). Ambient ionization mass spectrometry imaging of rohitukine, a chromone anticancer alkaloid, during seed development in *Dysoxylum binectariferum* Hook. f (Meliaceae). *Phytochemistry*, *116*, 104–110. doi:10.1016/j.phytochem.2015.02.031

Lane, A. L., Nyadong, L., Galhena, A. S., Shearer, T. L., Stout, E. P., Parry, R. M., ... Kubanek, J. (2009). Desorption electrospray ionization mass spectrometry reveals surface-mediated antifungal chemical defense of a tropical seaweed. *Proceedings of the National Academy of Sciences, 106*(18), 7314–7319. doi:10.1073/pnas.0812020106

Lewis, C. R., Carson, R. H., Erikson, M. N., Zagieboylo, A. P., Naylor, B. C., Li, K. W., ... Farnsworth, P. B. (n.d.). Desorption electrospray ionization (DESI) mass spectrometric imaging of spatially regulated in vivo metabolic rates. All Theses and Dissertations. 6555. https://scholarsarchive.byu.edu/etd/6555

Li, B., Bjarnholt, N., Hansen, S. H., & Janfelt, C. (2011). Characterization of barley leaf tissue using direct and indirect desorption electrospray ionization imaging mass spectrometry. *Journal of Mass Spectrometry, 46*(12), 1241–1246. doi:10.1002/jms.2010

Li, B., Hansen, S. H., & Janfelt, C. (2013). Direct imaging of plant metabolites in leaves and petals by desorption electrospray ionization mass spectrometry. *International Journal of Mass Spectrometry, 348*, 15–22. doi:10.1016/j.ijms.2013.04.018

Mohana Kumara, P., Srimany, A., Arunan, S., Ravikanth, G., Uma Shaanker, R., & Pradeep, T. (2016). Desorption electrospray ionization (DESI) mass spectrometric imaging of the distribution of rohitukine in the seedling of *Dysoxylum binectariferum* Hook. F. *PloS One, 11*(6), e0158099. doi:10.1371/journal.pone.0158099

Müller, T., Oradu, S., Ifa, D. R., Cooks, R. G., & Kr, B. (2011). Direct plant tissue analysis and imprint imaging by desorption electrospray ionization mass spectrometry. *Analytical Chemistry, 83*, 5754–5761. doi:10.1021/ac201123t

Perez, C. J., Tata, A., Campos, M. L. De, Peng, C., & Ifa, D. R. (2017). Monitoring toxic ionic liquids in Zebrafish (*Danio rerio*) with desorption electrospray ionization mass spectrometry imaging (DESI-MSI). *Journal of the American Society for Mass Spectrometry, 28*, 1136–1148. doi:10.1007/s13361-016-1515-9

Schwartz, S. A., Reyzer, M. L., & Caprioli, R. M. (2003). Direct tissue analysis using matrix-assisted laser desorption/ionization mass spectrometry: Practical aspects of sample preparation. *Journal of Mass Spectrometry, 38*(7), 699–708. doi:10.1002/jms.505

Sica, V. P., Raja, H. A., El-Elimat, T., & Oberlies, N. H. (2014). Mass spectrometry imaging of secondary metabolites directly on fungal cultures. *RSC Advances, 4*(108), 63221–63227. doi:10.1039/c4ra11564c

Takáts, Z., Wiseman, J. M., & Cooks, R. G. (2005). Ambient mass spectrometry using desorption electrospray ionization (DESI): Instrumentation, mechanisms and applications in forensics, chemistry, and biology. *Journal of Mass Spectrometry, 40*(10), 1261–1275. doi:10.1002/jms.922

Takáts, Z., Wiseman, J. M., Gologan, B., & Graham Cooks, R. (2004). Mass spectrometry sampling under ambient conditions with desorption electrospray ionization. *Science, 306*, 471–473. Retrieved from www.sciencemag.org/cgi/content/full/306/5695/469/www.sciencemag.orghttp://science.sciencemag.org/

Talaty, N., Takats, Z., & Cooks, R. G. (2005). Rapid in situ detection of alkaloids in plant tissue under ambient conditions using desorption electrospray ionization. *The Analyst, 130*, 1624–1633. doi:10.1039/b511161g

Thunig, J., Hansen, S. H., & Janfelt, C. (2011). Analysis of secondary plant metabolites by indirect desorption electrospray ionization imaging mass spectrometry. *Analytical Chemistry, 83*, 3256–3259.

Towers, M. W., Karancsi, T., Jones, E. A., Pringle, S. D., & Claude, E. (2018). Optimised desorption electrospray ionisation mass spectrometry imaging (DESI-MSI) for the analysis of proteins/peptides directly from tissue sections on a travelling wave ion mobility Q-ToF. *Journal of the American Society for Mass Spectrometry*, *29*(12), 2456–2466. doi:10.1007/s13361-018-2049-0

Veličković, D., & Anderton, C. R. (2017). Mass spectrometry imaging: Towards mapping the elemental and molecular composition of the rhizosphere. *Rhizosphere*, *3*(February), 254–258. doi:10.1016/j.rhisph.2017.03.003

Wu, C., Dill, A. L., Eberlin, L. S., Cooks, R. G., & Ifa, D. R. (2012). Mass spectrometry imaging under ambient conditions. *Mass Spectrometry Reviews*, 32(3), 218–243. doi:10.1002/mas

CHAPTER 7

Liquid Extraction Surface Analysis

Leo M.L. Nollet

CONTENTS

7.1 TECHNIQUE

Liquid extraction surface analysis (LESA) is a surface extraction method via a liquid droplet held in suspension (liquid junction) allowing a novel approach for spatially resolved analysis of biological and technical surfaces of interest. It combines micro-liquid extraction from a solid surface with nano-electrospray mass spectrometry (1).

LESA is a new, high-throughput tool for ambient mass spectrometry. A solvent droplet is deposited from a pipette tip onto a surface and maintains contact with both the surface and the pipette tip for a few seconds before being re-aspirated. The technique is particularly suited to the analysis of trace materials on surfaces due to its high sensitivity and low volume of sample removal (Figure 7.1).

LESA is a commercial variant of the liquid microjunction surface sampling probe (LMJ SSP) developed by Van Berkel (2) and made available on the Triversa Nanomate nanoelectrospray platform (3). The LESA extraction process is very simple: a droplet of solvent is applied to the surface of a substrate by a robotically controlled pipette tip. The droplet is held for a few seconds, forming a liquid microjunction between the pipette tip and the surface and allowing the passive diffusion of analyte molecules from the surface and into the solvent. The sample is then re-aspirated and infused into the mass spectrometer via ESI. The approach is suitable for a range of different analytes, including small molecules, lipids, proteolytic peptides, and intact proteins.

LESA-MS is found to be a robust, fast, and discriminating approach.

7.2 APPLICATIONS

Liquid extraction surface analysis, coupled with infusion nano-electrospray high-resolution mass spectrometry and tandem mass spectrometry (MS/MS), is described and applied by Eikel and Henion (4) to the qualitative determination of surface chemical residues resulting from the artificial spraying of selected fresh fruits and vegetables with

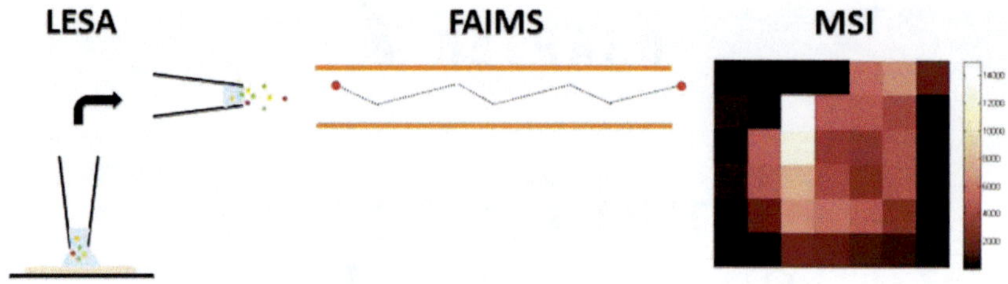

FIGURE 7.1 Scheme of LESA-MSI. FAIMS: field asymmetric waveform ion mobility spectrometry. (From *Anal. Chem.* 2016, 88, 13, 6758–6766.)

representative pesticides. Each of the targeted pesticides was readily detected with both high-resolution and full-scan collision-induced dissociation (CID) mass spectra. In the case of simazine and sevin, a mass resolution of 100,000 was insufficient to distinguish the isobaric protonated molecules for these compounds. When the surface of a spinach leaf was analyzed by LESA, trace levels of diazinon were readily detected on the spinach purchased directly from a supermarket before they were sprayed with the five-pesticide mixture. A 30 s rinse under hot running tap water appeared to quantitatively remove all remaining residues of this pesticide. Diazinon was readily detected by LESA analysis on the skin of the artificially sprayed spinach. Finally, incurred pyrimethanil at a level of 169 ppb in a batch slurry of homogenized apples was analyzed by LESA, and this pesticide was readily detected by both high-resolution mass spectrometry and full-scan CID mass spectrometry, thus showing that pesticides may also be detected in whole fruit homogenized samples. This report shows that representative pesticides on fruit and vegetable surfaces present at levels 20-fold below generally allowed EPA tolerance levels are readily detected and confirmed by the title technologies, making LESA-MS an interesting screening method for food safety purposes.

The detection and identification of proteins adsorbed onto biomaterial surfaces under ambient conditions has significant experimental advantages but has proven to be difficult to achieve with conventional measuring technologies. Rao et al. (5) present an adaptation of desorption electrospray ionization (DESI) and liquid extraction surface analysis (LESA) mass spectrometry (MS) coupled with in-situ surface tryptic digestion to identify protein species from a biomaterial surface. Cytochrome c, myoglobin, and BSA in a combination of single and mixture spots were printed in an array format onto Permanox slides, followed by in-situ surface digestion and detection via MS. Automated tandem MS performed on surface peptides was able to identify the proteins via MASCOT. A comparison of DESI and LESA-MS peptide spectra characteristics and sensitivity was made. DESI-MS images of the arrays were produced and analyzed with imaging multivariate analysis to automatically separate peptide peaks for each of the proteins within a mixture into distinct components. This is the first time that DESI and LESA-MS have been used for the in-situ detection of surface digested proteins on biomaterial surfaces and presents a promising proof of concept for the use of ambient MS in the rapid and automated analysis of surface proteins.

The suitability of LESA for obtaining detailed chemical profiles of fingerprints, oral fluid, and urine, which may be used in future for rapid medical diagnostics or metabolomics studies, was assessed (6). It was shown that LESA can be used to detect illicit drugs and their metabolites in urine, oral fluid, and fingerprints. This makes LESA a potentially useful tool in the growing field of fingerprint chemical analysis. This can be relevant not only to forensics but also to medical diagnostics. LESA can also be used to detect the explosive material RDX in contaminated artificial fingermarks.

The rapid and direct analysis of the amount and spatial distribution of exogenous chloroquine (CHQ) and CHQ metabolites from tissue sections by LESA-MS/MS was demonstrated by Parson et al. (7). LESA-MS/MS results compared well with previously published CHQ quantification data collected by organ excision, extraction, and fluorescent detection. The ability to directly sample and analyze spatially resolved exogenous molecules from tissue sections with minimal sample preparation and analytical method development has the potential to facilitate the assessment of target tissue penetration of pharmaceutical compounds, to establish pharmacokinetic/pharmacodynamic relationships, and to complement established pharmacokinetic methods used in the drug discovery process during tissue distribution assessment.

Top-down identification of proteins by LESA mass spectrometry has previously been reported for tissue sections and dried blood spot samples. Randall et al. (8) present a modified "contact" LESA method for top-down analysis of proteins directly from living bacterial colonies grown in Petri dishes, without any sample pretreatment. It was possible to identify a number of proteins by the use of collision-induced dissociation tandem mass spectrometry followed by searches of the data against an *E. coli* protein database. The proteins identified suggest that the method may provide insight into the bacterial response to environmental conditions. Moreover, the results show that the "contact" LESA approach results in a smaller sampling area than typical LESA, which may have implications for spatial profiling.

LESA-MS methodology is implemented to analyze five types of thermally treated meat species, namely, beef, pork, horse, chicken, and turkey meat, to select and identify heat-stable and species-specific peptide markers (9). In-solution tryptic digests of cooked meats were deposited onto a polymer surface, followed by LESA-MS analysis and evaluation using multivariate data analysis and tandem electrospray MS. The five types of cooked meat were clearly discriminated using principal component analysis and orthogonal partial least-squares discriminant analysis. Twenty-three heat-stable peptide markers unique to species and muscle protein were identified following data-dependent tandem LESA-MS analysis. Surface extraction and direct ambient MS analysis of mixtures of cooked meat species was performed for the first time and enabled the detection of 10% (w/w) of pork, horse, and turkey meat and 5% (w/w) of chicken meat in beef, using the developed LESA-MS/MS analysis. The study shows, for the first time, that ambient LESA-MS methodology displays specificity sufficient to be implemented effectively for the analysis of processed and complex peptide digests. The proposed approach is much faster and simpler than other measurement tools for meat speciation; it has potential for application in other areas of meat science or food production.

The use of ambient desorption electrospray ionization mass spectrometry (DESI-MS) and LESA-MS is explored for the first time to analyze skeletal muscle proteins obtained from a mixture of standard proteins and raw meat (10). Single proteins and mixtures of up to five proteins (myoglobin, troponin C, actin, bovine serum albumin (BSA), tropomyosin) were deposited onto a polymer surface, followed by in-situ tryptic digestion and comparative analysis using DESI-MS and LESA-MS using tandem electrospray MS. Peptide peaks specific to individual proteins were readily distinguishable with a good signal-to-noise ratio in the five-component mixture. LESA-MS gave a more stable analysis and greater sensitivity compared with DESI-MS. Meat tryptic digests were subjected to peptidomics analysis by DESI-MS and LESA-MS. Bovine, horse, pig, chicken, and turkey muscle digests were clearly discriminated using multivariate data analysis (MVA) of the peptidomic data sets. The most abundant skeletal muscle proteins were identified and correctly classified according to the species following MS/MS analysis.

REFERENCES

1. N. J. Martin, R. L. Griffiths, R. L. Edwards, and H. J. Cooper. Native liquid extraction surface analysis mass spectrometry: Analysis of noncovalent protein complexes directly from dried substrates. *Journal of the American Society for Mass Spectrometry*, 26 (8), 1320–1327, 2015.
2. G. J. Van Berkel and V. Kertesz. Continuous-flow liquid microjunction surface sampling probe connected on-line with high-performance liquid chromatography/mass spectrometry for spatially resolved analysis of small molecules and proteins. *Rapid Communications in Mass Spectrometry*, 27 (12), 1329–1334, 2013.
3. https://advion.com/products/triversa-nanomate/
4. D. Eikel and J. Henion. Liquid extraction surface analysis (LESA) of food surfaces employing chip-based nano-electrospray mass spectrometry. *Rapid Communications in Mass Spectrometry*, 25 (16), 2345–2354, 2011.
5. Wei Rao, Adam D. Celiz, David J. Scurr, Morgan R. Alexander, and David A. Barrett. Ambient DESI and LESA-MS analysis of proteins adsorbed to a biomaterial surface using in-situ surface tryptic digestion. *Journal of the American Society for Mass Spectrometry*, 24,1927Y1936, 2013.
6. M. Bailey, E. C. Randall, C. Costa, T. L. Salter, A. M. Race, M. de Puit, M. Koeberg, M. Baumert, and J. Bunch. Analysis of urine, oral fluid and fingerprints by liquid extraction surface analysis coupled to high resolution MS and MS/MS – Opportunities for forensic and biomedical science. *Analytical Methods*, 16, 3373–3382, 2016.
7. W. B. Parson, S. L. Koeniger, R. W. Johnson, J. Erickson, Y. Tian, C. Stedman, A. Schwartz, E. Tarcsa, R. Cole, and G. J. Van Berkel. Analysis of chloroquine and metabolites directly from whole-body animal tissue sections by liquid extraction surface analysis (LESA) and tandem mass spectrometry. *Journal of Mass Spectrometry*, 47 (11), 1420–1428, 2012.
8. E. C. Randall, J. Bunch, and H. J. Cooper. Direct analysis of intact proteins from *Escherichia coli* colonies by liquid extraction surface analysis mass spectrometry. *Analytical Chemistry*, 86 (21), 10504–10510, 2014.
9. M. Montowska, M. R. Alexander, G. A. Tucker, and D. A. Barrett. Rapid detection of peptide markers for authentication purposes in raw and cooked meat using ambient liquid extraction surface analysis mass spectrometry. *Analytical Chemistry*, 86 (20), 10257–10265, 2014.
10. M. Montowska, W. Rao, M. R. Alexander, G. A. Tucker, and D. A. Barrett. Tryptic digestion coupled with ambient desorption electrospray ionization and liquid extraction surface analysis mass spectrometry enabling identification of skeletal muscle proteins in mixtures and distinguishing between beef, pork, horse, chicken, and turkey meat. *Analytical Chemistry*, 86 (9), 4479–4487, 2014.

CHAPTER 8

Laser Ablation Inductively Coupled Plasma Mass Spectrometry

Leo M.L. Nollet

CONTENTS

8.1 TECHNIQUE

Most of the information in following paragraphs of this section, *Technique*, are taken from the website of *Applied Spectra* Inc, West Sacramento, CA.

Laser ablation inductively coupled plasma mass spectrometry (LA-ICP-MS) is a powerful analytical technology enabling highly sensitive elemental and isotopic analysis directly on solid samples (1) (Figure 8.1).

LA-ICP-MS begins with a laser beam focused on the sample surface to generate fine particles. This process is known as laser ablation. It is a process in which a laser beam is focused on a sample surface to remove material from the irradiated zone (2). Laser ablation is a dynamic technology platform for performing chemical analysis. LA generates a bright plasma on the sample's surface. The light emitted from this plasma can be analyzed to determine the presence and concentration levels of elements in the periodic table; this technology is known as laser-induced breakdown spectroscopy (LIBS). LA also creates a flux of fine particulates that can be readily transported to a secondary source for ICP-MS analysis (LA-ICP-MS).

The ablated particles are then transported to the secondary excitation source of the ICP-MS instrument for digestion and ionization of the sampled mass. The excited ions in the plasma torch are subsequently introduced to a mass spectrometer detector for both elemental and isotopic analysis.

LA-ICP-MS can perform ultra-highly sensitive chemical analysis down to ppb level, without any sample preparation.

Samples can be both conducting or non-conducting, and the analysis can be performed in the air without the need for a complex vacuum system. Results are available within seconds. The sample mass size required for LA-ICP-MS analysis is in the order of picograms to femtograms.

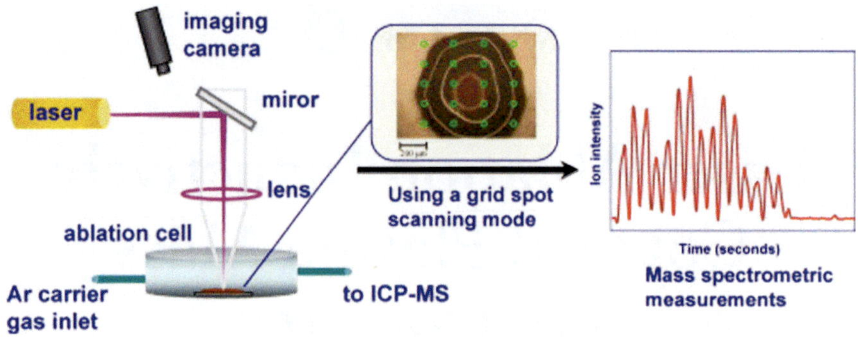

FIGURE 8.1 Setup of LA-ICP-MS.

The use of femtosecond laser pulsing represents the latest technical advance in LA-ICP-MS technology. It improves precision and accuracy, reduces the elemental/isotopic fractionation, and enhances the sensitivity of the measurements.

Laser ablation-based analytical techniques such as LIBS and LA-ICP-MS offer the ability to perform rapid chemical analysis directly from solid samples without the tedious and complex sample preparation involving toxic acid solutions.

Laser parameters (energy, wavelength, pulse duration) and sample physical/chemical properties as well as ambient gas will influence the various mechanisms of the ablation. They often govern optical/physical properties of the generated plasma, the primary and secondary mechanisms responsible for mass removal, and the particle size distribution of the ablated mass. Consequently, a proper selection of appropriate laser parameters based on the understanding of laser ablation fundamentals is a critical prerequisite for achieving accurate and precise analytical measurements using LIBS and LA-ICP-MS.

Only a few studies on LA-ICP-MS in food analysis have appeared in the literature. A greater number of papers on that subject deal with applications in pathophysiology, pharmacology, and toxicology.

8.2 APPLICATIONS IN BIOLOGICAL TISSUES

Here, the discussion starts with a number of LA-ICP-MS techniques in biological tissues.

J.S. Becker et al. (3) developed an analytical imaging technique using LA-ICP-MS to study element distribution in biological tissues. Nowadays, imaging LA-ICP-MS using double-focusing sector field (LA-ICP-SF-MS) or quadrupole-based mass spectrometers (LA-ICP-QMS) can be applied as an exciting tool providing new information on the pathophysiology, pharmacology, and toxicology of elements of interest in biological systems.

The quantitative determination of elements (e.g., Cu, Fe, Zn, Se, and others) in biological tissues is of growing interest especially in brain research (e.g., for studying neurodegenerative diseases such as Alzheimer's or Parkinson's disease). LA-ICP-SFMS was employed to produce images of detailed regionally specific element distributions in thin tissue sections of different sizes (such as control human or rat brain tissues or tumor regions).

In addition, imaging MS using LA-ICP-QMS was applied to study the uptake and transport of nutrient and toxic elements in plant tissues.

Besides the quantitative imaging of essential and toxic elements in tissues, powerful analytical techniques are also required for the determination and characterization of phosphoproteins and metal-containing proteins within a large pool of proteins, after electrophoretic separation [e.g., blue native polyacrylamide gel electrophoresis (BN-PAGE) and sodium dodecyl sulfate polyacrylamide gel electrophoresis (SDS-PAGE)] into 1D and 2D gels. LA-ICP-MS was used to detect metalloproteins in protein bands of 1D gels or protein spots separated after 2D gel electrophoresis (2D-GE). In addition to elemental determination by LA-ICP-MS, matrix-assisted laser desorption/ionization (MALDI-MS) was employed to identify metal-containing proteins.

The distribution analysis of (essential, beneficial, or toxic) metals (e.g., Cu, Fe, Zn, Pb, and others), metalloids, and non-metals in biological tissues is of key interest in life science. Especially, in brain research, metalloproteins are the focus of targeted therapy approaches of neurodegenerative diseases such as Alzheimer's and Parkinson's disease, or stroke, or tumor growth.

LA-ICP-MS using double-focusing sector field or quadrupole-based mass spectrometers has been successfully applied as a powerful imaging technique to produce quantitative images of detailed regionally specific element distributions in thin tissue sections of human or rodent brain (4). Imaging LA-ICP-QMS was also applied to investigate metal distributions in plant and animal sections to study, for example, the uptake and transport of nutrient and toxic elements or environmental contamination. The combination of imaging LA-ICP-MS of metals with proteomic studies using biomolecular mass spectrometry identifies metal-containing proteins and also phosphoproteins. Metal-containing proteins were imaged in a two-dimensional gel after electrophoretic separation of proteins (SDS or Blue Native PAGE). Recent progress in LA-ICP-MS imaging as a stand-alone technique and in combination with MALDI/ESI-MS for selected life science applications is summarized.

A novel and straightforward procedure for the preparation of matrix matched calibration standards for the quantitative imaging of iron (Fe) in biological tissues by laser ablation LA-ICP-MS with on-tissue internal standard addition was described (5). This simple approach enabled on-tissue addition of Rh as internal standard to samples (with heterogeneous Fe distribution) and calibrants (with homogeneous Fe distribution). This was achieved without altering the original Fe distribution of the sample.

A literature survey over the last ten years concerning the use of LA-ICP-MS for biological tissues analysis was reviewed by D. Pozebon et al. (6). Instrumentation, strategies of calibration for quantitative analysis, challenges, and recent advances in this field are discussed. Applications of isotope ratio (IR), including trace experiments, and isotope dilution (ID) are reviewed for biological samples. Bioimaging methods, studies, and applications to animal and plants tissues are emphasized, demonstrating the importance of images of metals and metalloids in biomedical research, and bioaccumulation and bioavailability studies for ecological and toxicological risk assessment in humans, animals, and plants. The usefulness of IR associated with bioimaging for predicting geographical origin, habitat, movement of subjects, diet, and lifestyle are also demonstrated.

The binding of Cu and Zn on bovine serum proteins was studied using a tracer experiment. One-dimensional BN-PAGE gels with separated bovine serum proteins after gel electrophoresis were doped with an enriched isotope copper tracer (^{65}Cu) as a function of time (from 30 s up to 24 h) (7). In several protein bands of the 1D gel before the tracer experiment, metal ions (mostly Zn) were found by LA-ICP-MS. After experiments using enriched ^{65}Cu tracer $^{65}Cu/^{63}Cu$ isotope ratios with changed isotope ratio were measured by LA-ICP-MS. The isotope ratio measurements of copper by LA-ICP-MS as a function of time indicate a fast exchange of Zn bonded on bovine serum albumin. This experimental finding demonstrates the formation of copper-binding proteins during the tracer experiments in the 1D gel. The identification of the protein bands was attempted from bovine serum proteins separated by 1D gel electrophoresis by using biopolymer mass spectrometry with MALDI-TOF-MS after excision from the 1D gel and tryptic digestion.

The review of A. Limbeck et al. (8) focuses on the application of LA-ICP-MS for the quantification of trace elements in environmental, biological, and medical samples. The fundamental problems of LA-ICP-MS, such as sample-dependent ablation behavior and elemental fractionation, can be even more pronounced in environmental and life science applications as a result of the large variety of sample types and conditions. Besides variations in composition, the range of available sample states is highly diverse, including powders (e.g., soil samples, fly ash), hard tissues (e.g., bones, teeth), soft tissues (e.g., plants, tissue thincuts), or liquid samples (e.g., whole blood). Within this article, quantification approaches that have been proposed in the past are critically discussed and compared regarding the results obtained in the applications described. Although a large variety of sample types is discussed, the quantification approaches used are similar for many analytical questions and have only been adapted to the specific questions. Nevertheless, none of them has proven to be a universally applicable method.

8.3 APPLICATIONS IN FOOD AND RELATED ITEMS

The quality and safety of food samples always require strict control. From an analytical perspective, many techniques can be used for this type of monitoring, one of which involves the application of laser pulses in LA-ICP-MS. One of the difficulties and challenges associated with this technique involves quantitative analyses because of problems in the process of ablation, plasma formation, and/or matrix effects. Calibration strategies combining internal standards, standard addition methods, and chemometric tools can help improve the results. Food samples are among the most variable samples in terms of their matrix and composition, which can further complicate analysis by laser ablation and require the development of ablation strategies. The study of dos Santos Augusto et al. describes the use of calibration strategies and the determination of Ca, Cd, Cr, Cu, Fe, Mg, Pb, and Zn in food samples, including liquid (i.e., orange juice) and solid (i.e., dietary supplements) samples (9). The calibration curve for the solid samples was constructed using microcrystalline cellulose and a proportional mixture of solid samples chosen according to their higher and lower concentrations of metals. The liquid samples were immobilized in a polymer film with the help of polyvinyl alcohol (PVA). Using these calibration strategies, it was possible to obtain accuracy values between 60 and 120% for almost all samples, except for Ca, Mg, and Zn. These exceptions could be a reflection of unresolved matrix interferences. Carbon was used as the internal standard but did not show promising results.

A protocol was validated for the elemental analysis of plant and food materials using LA-ICP-MS and pressed pellets in compliance with the Green Chemistry principles, i.e., avoiding the use of hazardous reagents (10). The protocol included sample pre-treatment, preparation of pellets, and LA-ICP-MS optimization for the food sample matrix using leaf certified reference materials. Sample homogenization and particle size reduction (~10 μm) for samples with larger initial particle sizes proved crucial in improving signal stability and analytical results, whereas the raster mode was preferred over the spot mode for greater accuracy. The validation of the technique was successful for determination of Mn, Ni, Cu, Zn, As, Se, Cd, and Pb (z-score values ≤ 2).

Coptis chinensis Franch., is a widely used medicinal plant in China. This plant is often contaminated by Cd. Understanding the distribution of Cd and its chemical forms is important to evaluate accumulation of the metal and its detoxification mechanisms in this plant. LA-ICP-MS was used to spatially locate Cd in rhizome cross-sections, and ICP-MS was used to analyze the Cd subcellular distribution and the chemical forms of Cd in different tissues (11). Rhizome bioimaging results showed that Cd was distributed predominantly within the periderm, cortex, pith, and root trace vascular bundle. The LA-ICP-MS results suggested that Ca^{2+} channels might be a pathway for Cd entry into the plant. Subcellular distribution data indicated that most of the Cd was associated with the cell wall (41.8–77.1%) and the soluble fraction (14.4–52.7%) in all tissues. Analysis of chemical forms revealed that the majority of the Cd existed in less mobile and less toxic forms in all tissues, and P could convert to insoluble phosphate with Cd to moderate Cd toxicity.

A LA-ICP-MS method was developed to determine Cu, Zn, Cd, Hg, and Pb and metal distribution in longitudinal tissue sections of the marine snail *Nassarius reticulatus* (Gastropoda, Prosobranchia) (12). Snails were sampled in northern Brittany (France) at three stations with different contamination levels. The quantification of metal distribution (imaging or mapping) in a thin slice of the snail tissue was carried out using different strategies: by one-point calibration and via matrix-matched laboratory standards using different biological materials (BCR 278, snail tissue, and rat brain). Together with the imaging of metals the distribution of two non-metals (carbon and sulfur) was analyzed. The imaging LA-ICP-MS analysis yielded an inhomogeneous distribution for all elements investigated. The detection limits for the distribution analysis of Cu, Zn, Cd, Hg, and Pb measured by LA-ICP-MS were in the low $\mu g\ g^{-1}$ range.

LA-ICP-MS was used for the quantitative imaging of nutrient elements (such as K, Mg, Mn, Cu, P, S, and B) in the leaves of *Elsholtzia splendens* (13). The plant leaves were scanned directly with a focused Nd:YAG laser in the laser ablation chamber. The ablated material was transported with argon as the carrier gas to a quadrupole-based ICP-MS (ICP-QMS), and the ion intensities of $^{39}K^+$, $^{24}Mg^+$, $^{55}Mn^+$, $^{63}Cu^+$, $^{31}P^+$, $^{34}S^+$, and $^{11}B^+$ were measured by ICP-QMS to study the distribution of the elements of interest. The imaging technique using LA-ICP-MS on plant leaves does not require any sample preparation. Carbon ($^{13}C^+$) was used as an internal standard element to compensate for the difference in the amount of material ablated.

Baba et al. (14) used LA-ICP-MS to determine Cd in rice flour, employing the responses from Cd-doped NIES rice flour standards 10a, 10b, and 10c to construct calibration curves. Good linearity was observed in the range 0.023–1.82 mg g^{-1}. Internal standardization was unnecessary, and no significant interferences were observed. Such studies show the typical advantages of LA-ICP-MS of no sample preparation and rapid sample throughput, which might be more widely employed in quality control of foodstuffs during or after processing.

The provenance identification of selected types of food and drink was undertaken from data obtained using both solution and laser ablation-based inductively coupled plasma mass spectrometry (ICP-MS) and solution-based inductively coupled plasma atomic emission spectroscopy (ICP-AES) analysis of selected food and drink materials (pork, wine, tea, coffee, and olive oil) (15). Quantitative determination of up to 55 elements was undertaken using solution-based ICP-MS and AES while counts-per-second data for 49 elements were used for samples analyzed using LA-ICP-MS. ICP-AES was used to determine the concentrations of Na, K, Ca, Mg, and Fe where appropriate. A quantitative analytical methodology for LA-ICP-MS analysis of olive oil was also developed and preliminary trials undertaken. Additional information, to establish more detailed geographic resolution in provenance determination, especially for tea and coffee (plantation of origin), was provided by the incorporation of data for light-stable isotope distribution patterns (2 H (D), 13C, and 15N). LA-ICP-MS was used for the direct analysis of olive oil and coffee beans. The use of laser ablation for the analysis of olive oil has the potential to overcome problems associated with both acid dissolution and thermal oxidation of this material, which are the traditional methods of sample dissolution and which either introduce significant contamination or cause the loss of volatile organometallic compounds that may be indicative for provenance determination. Results from the overall study indicated that it was possible not only to establish country of origin of the study materials but also in some cases to improve resolution of provenance to state of origin (wine and pork), and even plantation of origin (tea and coffee).

A method has been developed for a rapid and precise location of selenium-containing proteins in large two-dimensional (2D) electrophoresis gels (16). A sample was divided into four aliquots which were analyzed in parallel by 1D isoelectric focusing electrophoresis (IEF)–laser ablation (LA) inductively coupled plasma mass spectrometry (ICP-MS), 1D sodium dodecyl sulfate polyacrylamide gel electrophoresis (SDS PAGE)–LA-ICP-MS, and, in duplicate, by 2D IEF-PAGE. On the basis of the 1D electropherograms obtained, areas supposed to contain the largest concentrations of Se were subjected to LA-ICP-MS imaging to locate precisely the position of Se-containing proteins which were then identified in the parallel 2D gel by electrospray Orbitrap MS/MS. The method was applied to the identification and semiquantitative determination of selenium storage proteins in wheat.

Cadmium is one of the most widespread metals which is highly toxic, and can enter the human food chain directly *via* accumulation in nutritional plants. To better understand the transport and protein-binding capabilities of heavy metals cadmium uptake in *Spinacia oleracea* L., a typical nutritional plant with a high cadmium uptake susceptibility was investigated (17). A new strategy for the screening of cadmium-containing proteins was elaborated which consists of the extraction of proteins by application of an ultrasonic homogenizer and separation of the metal-binding proteins by use of native PAGE. For comparison, denaturing PAGE has been investigated, too. LA was utilized as a sample introduction system for ICP-MS detection of the metals in the proteins after PAGE separation and blotting onto membranes.

Concentrations of 12 elements (Mg, Al, Ca, V, Cr, Mn, Fe, Co, Ni, Cu, Zn, Cd) were determined in six herbal supplements (18), Korean Panax ginseng (*Panax ginseng*), goldenseal (*Hydrastis canadensis*), ginger root (*Zingiber officinale*), St. John's wort (*Hypericum perforatum*), green tea (*Camellia sinensis*), and valerian root (*Valeriana officinalis*), by both LA-ICP-MS and conventional closed-vessel digestion solution nebulization-ICP-MS (SN-ICP-MS). For LA-ICP-MS, powder from supplement capsules and leaf reference materials were pressed into pellets, the latter being used for calibration and quality assurance. Laser ablation was performed using line scans with a scan rate of 30 $\mu m\ min^{-1}$, a frequency of 20 Hz, and a spot size of 100 μm; ^{13}C served as the internal standard. For LA it was found that low resolution ($m/\Delta m \approx 400$) yielded good recoveries for the reference materials and results comparable to SN-ICP-MS for most elements, except for Ca, which was better determined in medium resolution ($m/\Delta m \approx 4000$).

With rice being the main staple crop in Asian countries such as China, Korea, and Japan, the detection of arsenic (As), an element known to be carcinogenic to humans, has been the topic of high public interest. The total arsenic content in 200 white and 104 brown rice samples collected in Korea was analyzed using a quadrupole inductively coupled plasma-mass spectrometer (ICP-MS) (19). One of the rice grain samples was polished with 3, 5, 7, 9, and 11 degrees of milling, and arsenic concentration variance from the surface to the inner core region was investigated. Furthermore, spatial distribution of arsenic over the cross-section of a brown rice grain was obtained using femtosecond laser ablation-inductively coupled plasma-mass spectrometry (fs LA-ICP-MS). For the total arsenic content analysis, 91.7 ± 28.1 and 101 ± 33.6 μg^{-1} kg^{-1} of arsenic were measured in the white and the brown rice, respectively. The fs LA-ICP-MS mapping image explains that the higher arsenic concentration in the brown rice is due to high arsenic distribution in the rice husk (protective covering of rice). Consequently, some degree of rice milling may be effective in the reduction of arsenic intake.

REFERENCES

1. https://appliedspectra.com/technology/la-icp-ms.html
2. https://appliedspectra.com/technology/laser-ablation.html
3. J. S. Backer, M. Zoriy, V. L. Dressler, B. Wul, J. S. Becker. Imaging of metals and metal-containing species in biological tissues and on gels by laser ablation inductively coupled plasma mass spectrometry (LA-ICP-MS): A new analytical strategy for applications in life sciences. *Pure and Applied Chemistry*, 80 (12), 2643–2655, 2008.
4. J. S. Becker, M. Zoriy, A. Matusch, B. Wu, D. Salber, C. Palm, J. S. Becker. Bioimaging of metals by laser ablation inductively coupled plasma mass spectrometry (LA-ICP-MS). *Mass Spectrometry Reviews*, 29, 156–175, 2010.
5. J. O'Reilly, D. Douglas, J. Braybrook, P.-W. So, E. Vergucht, J. Garrevoet, B. Vekemans, L. Vincze, H. Goenaga-Infante. A novel calibration strategy for the quantitative imaging of iron in biological tissues by LA-ICP-MS using matrix-matched standards and internal standardization. *Journal of Analytical Atomic Spectrometry*, 29, 1378–1384, 2014.
6. D. Pozebon, L. G. L. Sheffler, V. Dressler, M. A. G. Nunes. Review of biological samples analysis using laser ablation inductively coupled plasma mass spectrometry (LA-ICP-MS). *Journal of Analytical Atomic Spectrometry*, 29, 2204–2228, 2014.
7. J. S. Becker, D. Pozebon, V. L. Dressler, R. Lobinskia, J. S. Becker. LA-ICP-MS studies of zinc exchange by copper in bovine serum albumin using an isotopic enriched copper tracer. *Journal of Analytical Atomic Spectrometry*, 23, 1076–1082, 2008.
8. A. Limbeck, P. Galler, M. Bonta, G. Bauer, W. Nischkauer, F. Vanhaecke. Recent advances in quantitative LA-ICP-MS analysis: Challenges and solutions in the life sciences and environmental chemistry. *Analytical and Bioanalytical Chemistry*, 407, 6593–6617, 2015.
9. A. dos Santos Augusto, M. A. Sperança, D. F. Andrade, E. R. Pereira-Filho. Nutrient and contaminant quantification in solid and liquid food samples using laser-ablation inductively coupled plasma-mass spectrometry (LA-ICP-MS): Discussion of calibration strategies. *Food Analytical Methods*, 10, 1515–1522, 2017.
10. E. M. Papaslioti, A. Parviainen, M. J. Román Alpiste, C. Marchesi, C. J. Garrido. Quantification of potentially toxic elements in food material by laser ablation-inductively coupled plasma-mass spectrometry (LA-ICP-MS) via pressed pellets. *Food Chemistry*, 274, 726–732, 2019.
11. W. Huang, Z. Bai, J. Jiao, H. Yuan,Z. Bao, S. Chen, M. Ding, Z. Liang. Distribution and chemical forms of cadmium in *Coptis chinensis* Franch. determined by laser ablation ICP-MS, cell fractionation, and sequential extraction. *Ecotoxicology and Environmental Safety*, 171, 894–903, 2019.

12. M. C. Santos, M. Wagner, B. Wu, J. Scheider, J. Oehlman, S. Cadore, J. S. Becker. Biomonitoring of metal contamination in a marine prosobranch snail (*Nassarius reticulatus*) by imaging laser ablation inductively coupled plasma mass spectrometry (LA-ICP-MS). *Talanta*, 80 (2), 428–433, 2009.
13. B. Wu, M. Zoriy, Y. Chen, J. S. Becke. Imaging of nutrient elements in the leaves of Elsholtzia splendens by laser ablation inductively coupled plasma mass spectrometry (LA-ICP-MS). *Talanta*, 78 (1), 132–137, 2009.
14. K. Baba, E. Watanabe, H. Eun and M. Ishizaka, Direct determination of cadmium in rice flour by laser ablation-ICP-MS. *Journal of Analytical Atomic Spectrometry*, 18 (12), 1485–1488, 2003.
15. R. J. N Waiting, G. S. Lee, C. J. Scadding, T. S. Pilgrim, R. L. Green, A. E. Martin, C. D. May, J. L. Valentin. The application of solution and laser ablation based ICP-MS and solution based AES for the provenance determination of selected food and drink produce. *The Open Chemical and Biomedical Methods Journal*, 3, 179–196, 2010.
16. J. Bianga, E. Govasmark, J. Szpunar. Characterization of selenium incorporation into wheat proteins by two-dimensional gel electrophoresis–Laser ablation ICP MS followed by capillary HPLC–ICP MS and electrospray linear trap quadrupole orbitrap MS. *Analytical Chemistry*, 85 (4), 2037–2043, 2013.
17. A. Polatajko, M. Azzolini, I. Feldmann, T. Stuezel, N. Jakubowski. Laser ablation-ICP-MS assay development for detecting Cd- and Zn-binding proteins in Cd-exposed Spinacia oleracea L. *Journal of Analytical Atomic Spectrometry*, 22, 878–887, 2007.
18. K. Bu, J. V. Cizdziel, L. Reidy. Analysis of herbal supplements for selected dietary minerals and trace elements by laser ablation- and solution-based ICPMS. *Microchemical Journal*, 106, 244–249, 2013.
19. S. H. Choi, J. S. Kim, J. Y. Lee, J. S. Jeon, J. W. Kim, R. E. Russo, J. Gonzalez, J. H. Yoo, K. S. Kim, J. S. Yang, K. S. Park. Analysis of arsenic in rice grains using ICP-MS and fs LA-ICP-MS. *Journal of Analytical Atomic Spectrometry*, 29, 1233–1237, 2014.

CHAPTER 9

Different Mass Spectrometry Imaging Methods

Sadaf Jamal Gilani, Chandrakala, and Md. Taleu Zzaman

CONTENTS

9.1 INTRODUCTION

Mass spectrometry imaging (MSI) has become evident in the last decade as a significant tool attributable to its distinctive potential to obtain molecularly specific images and a lot of information irrespective of the need for labeling or staining (Wu et al., 2013). Indisputably mass spectrometry (MS) contributes significantly to chemical, biological, and geological research (Nemes and Vertes, 2007). For instance, there is an unavailability of a single instrument platform for overall coverage of plant metabolites. In contrast to other techniques, MSI warrants label-free, high-resolution spatial mapping of a wide range of biomolecules inclusive of its qualitative and quantitative chemical information, in a single experiment (Kulkarni et al., 2018). This chapter discusses recent developments and applications of MSI, emphasizing the ionization techniques of laser ablation electrospray ionization (LAESI), laser desorption ionization (LDI), desorption atmospheric pressure photoionization (DAPPI), desorption atmospheric chemical photoionization (DAPCI), low-temperature plasma (LTP), and laser ablation atmospheric pressure photoionization (LAAPPI).

9.2 LASER ABLATION ELECTROSPRAY IONIZATION (LAESI)

LAESI was developed in 2007 by Akos Vertes and Peter Nemes (Vertes and Nemes, 2007). It is a hybrid ambient ionization method involving laser desorption of the analyte followed by its ionization in the course of collisions of the ablated material with charged droplets produced by ESI (Wu et al., 2013). LAESI serves as a novel ionization source utilized by mass spectrometry and is used to perform MS imaging of plants, tissues, cell pellets, and even single cells.

9.2.1 Theory

LAESI is a hybrid ambient ionization method; it is a novel combination of representative features of MALDI and ESI. This technique utilizes a mid-infrared laser of wavelength of 2.94 μm that significantly ablates samples by absorbing water, resulting in the plume which is rich in the neutral molecules from samples (Figure 9.1). Ionization of the sample molecules is done by combining the sample molecule with a plume of electrospray followed by the passage of sample ion into the mass spectrometer for detection. The significance of this direct analysis is the alleviation of sample contamination or loss as no sample preparation or pretreatment is required. 2D and 3D imaging of cell colonies, plant and tissue sections, and contact lenses are possible on the usage of the natural water content of the tissue. A variety of molecular classes and size ranges of up to 66 kDa can be detected with this technique. (Nemes and Vertes, 2007). The samples are ablated using an Er: YAG laser at 90° incident angle after mounting it on microscope slides, 10–30 mm below the spray axis and 3–5 mm in front of the emitter tip. The water in the sample acts as an intrinsic matrix; the OH vibration of the water molecule is excited by a focused mid-infrared laser beam of 2940 nm. The ablated craters are of diameters typically between 200–400 μm and 40–50 μm in depth. (Nemes and Vertes, 2007). The similarity of LAESI, electrospray laser desorption ionization (ELDI), and laser electrospray MS (LEMS) is that these techniques rely on ESI post-ionization of neutrals from a laser plume. However, LAESI and ELDI are resonant processes due to

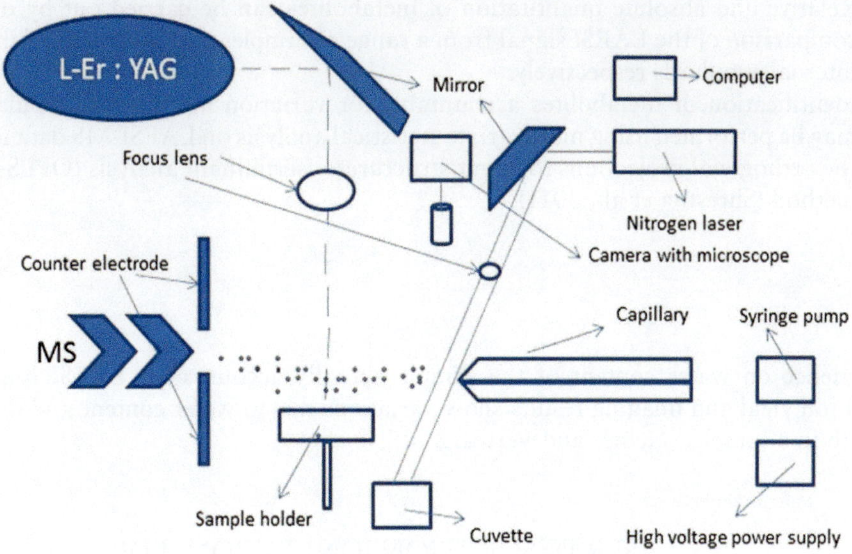

FIGURE 9.1 Schematic diagram of the LAESI setup (L-Er:YAG laser).

the use of nanosecond mid-IR and UV lasers; it is unlikely for LEMS to produce non-resonant laser ablation of the sample due to the use of a femtosecond near-IR laser (Wu et al., 2013).

9.2.2 Applications

Vertes and coworkers introduced applications of LAESI in 2008 (Nemes et al., 2008). They investigated rat brain sections by mass spectrometric analysis (Nemes et al., 2010).

1. LAESI was utilized for cell-by-cell imaging of plant tissues (*in situ*). With a lateral resolution of spot areas of 30 μm imaging of onions (*Allium cepa*), single epidermal cells and clusters of three to four achlorophyllous cells in citrus leaves (*Citrus aurantium*) was successful (Wu et al., 2013).
2. Direct analysis of body fluid such as urine, blood, and serum revealed antihistamine excretion. Identification of proteins, lipids, and metabolites was done effectively (Nemes and Vertes, 2007).
3. Metabolites on leaf, stem, and root in French marigold (*Tagetes patula*) were investigated *in vivo* spatial profiling (Nemes and Vertes, 2007).

9.2.3 Advantages

1. With the LAESI technique, sampling of the entire volume of cells for metabolites and lipid components is possible.
2. Simultaneous detection of ranges of the compounds such as acidic and basic compounds, lipids, and fatty acids is possible.

3. Relative and absolute quantitation of metabolites can be carried out by direct comparison of the LAESI signal from a range of samples and by utilizing labeled internal standards respectively.
4. Identification of metabolites accountable for variation among cell population may be performed using multivariate statistical analysis on LAESI-MS data using the orthogonal projections to latent structures discriminant analysis (OPLS-DA) method (Shrestha et al., 2011).

9.2.4 Limitations

Dependence on water content of the sample actually accounts for LAESI limitation. LAESI ion yield and imaging results show variations due to water content and/or tensile strength in a sample. (Nemes and Vertes, 2007)

9.3 LASER DESORPTION IONIZATION (LDI)

9.3.1 Theory

The laser desorption/ionization imaging method provides infinitesimal and simple sample preparation. This method is capable of detecting adulterations as it provides specific markers that characterize adulterations that can be used as proof of concept to identify one of the most adulterated drug products, sildenafil (Falco et al., 2019). Direct and semi-quantitative analysis of the sample is possible (Figure 9.2).

Matrix-assisted laser desorption (MALDI) is an extensively used vacuum MSI method. Imaging of large molecules, for instance, proteins, with spatial resolution extending out

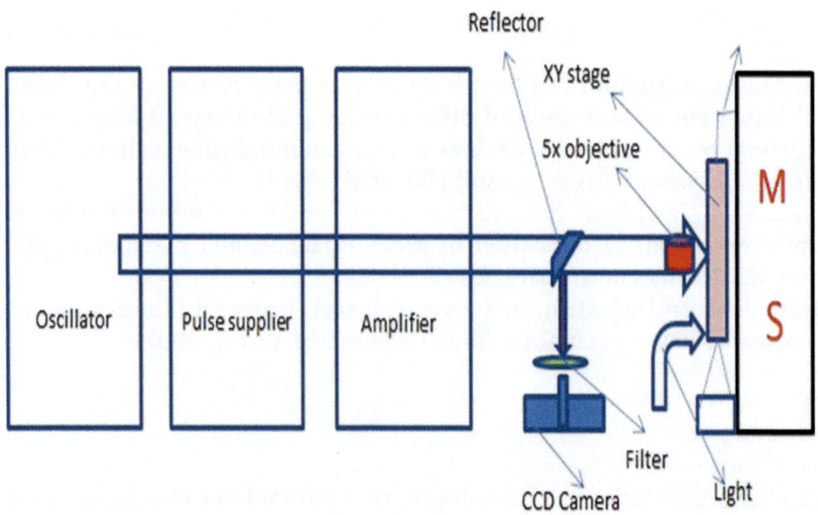

FIGURE 9.2 Schematic diagram of the atmospheric pressure femtosecond laser desorption ionization imaging mass spectrometer (AP fs-LDI IMS).

into tens of the micron-scale is possible with MALDI. Atmospheric pressure femtosecond laser desorption (AP fs-LDI MSI) uses focused NIR fem to second laser pulses devoid of any laser matrix for sample ablation and analyte. This technique offers the highest lateral resolution ever proclaimed for an ambient ionization technique. The lateral resolution of 10 μm reported for onion epidermis is a significant improvement and a step forward towards chemical visualization of individual cells in ambient MSI. Moreover, this short-pulse technique is advantageous as most plant and animal tissue sustains the analysis because of its superficiality. Furthermore, this opens up the feasibility of *in-vivo* chemical imaging (Mugo and Bottaro, 2007)

9.3.2 Advantages

1. Chemical imaging at a cellular level at atmospheric pressure is possible via this technique.
2. Analysis of low molecular weight organic molecules is feasible when nanomaterials are used as DI enhancers in laser desorption ionization mass spectrometry (Picca et al., 2017).
3. This technique is also applicable to understanding spatiotemporal aspects of biology via mapping of molecular distributions in basic biologic research. Better diagnosis and prognosis of disease can also be achieved for clinically relevant treatment (Caprioli, 2019).

9.3.3 Limitations

1. The background noise of matrix-related ions imposes difficulty in the detection of low molecular weight compounds (to approximately 700 *m/z*).
2. The absence of a universal matrix implies a preliminary knowledge of the investigated compounds.
3. Inhomogeneous matrix-analyte co-crystallization make the quantitative analysis difficult.

9.3.4 Applications

1. High molecular weight analytes such as protein and other macromolecules can be detected and characterized using MALDI (Mandal et al., 2019).
2. LDI-MS is suitable for the rapid chemical fingerprinting of lichen extracts (Pogam et al., 2015).
3. Proteins isolated by using gel electrophoresis can be rapidly identified by MALDI (Huberty et al., 1993)
4. MALDI can suitably determine the molar mass distribution in polymer chemistry (Schrepp and Pasch, 2003).
5. Medical microbiological laboratories utilize MALDI as a standard method for species identification (Seng et al., 2009).
6. MALDI can serve as an early detection technique and for the identification of protein associated with pancreatic cancer (Zhong et al., 2014).

9.4 DESORPTION ATMOSPHERIC PRESSURE PHOTOIONIZATION (DAPPI)

9.4.1 Theory

DAPPI was introduced by Haapala et al. in 2007 (Haapala et al., 2007). The unique characteristic of this technique is that it includes neutral and nonpolar compounds. In this technique, a hot mist of vaporized solvent such as toluene or acetone is delivered using a heated nebulizer microchip for sample desorption. A photoionization lamp (10 eV) is used as an ionization agent. Thermal desorption of the analyte is promoted by delivery of a heated jet of mixed nebulizer gas and spray solvent vapor to the sample (Figure 9.3). Surface temperatures up to 350°C have been reported. Photoionization of the sample then takes place by the classic gas-phase mechanism. Depending upon the spray solvent, analyte, and ionization mode employed, the initial photoionization products are converted via ion/molecule reactions to [M+H]+, [M−H]−, M+•, M−•, [M−H+O]−, or [M−2H+2O]− ions (Luosujarvi et al., 2008).

9.4.2 Applications

The application of DAPPI was first demonstrated by Pol et al. (Pol, et al., 2009).

1. The DAPPI set-up was used to analyze brain tissue. As brain tissue is rich in phospholipids content dominant signals could be seen at high resolution. This was also seen in the dehydrated leaf of *Salvia* (sage) (Pol et al., 2009).
2. The maximum spatial resolution was limited to approximately 1 mm due to the use of heated nebulizer microchip and the glass surface in the experiments. By performing DAPPI-MS on sage leaf, chemical species such as tocopherol (seen as the radical cation, *m/z* 430), carnosol (as the protonated molecule, *m/z* 331), and methyl carnosic acid (*m/z* 301) were mapped (Pol et al., 2009).

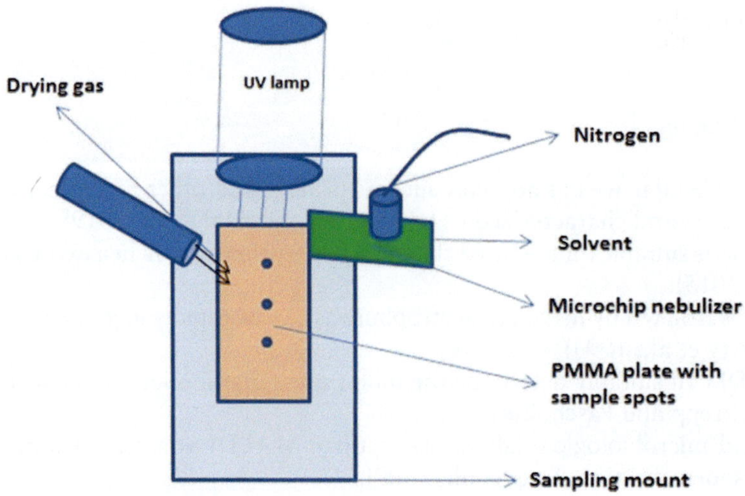

FIGURE 9.3 Schematic diagram of the DAPPI setup.

3. Cholesterol fragment for the human brain at mass 369 was the only peak in the spectrum observed by DAPPI using acetone as a dopant (Pol et al., 2009).

9.4.3 Limitation

A thermally degradable sample cannot be analyzed by this technique.

9.5 DESORPTION ATMOSPHERIC PRESSURE CHEMICAL IONIZATION (DAPCI)

9.5.1 Theory

Desorption atmospheric pressure chemical ionization (DAPCI) was first reported for the trace level detection of trinitrotoluene (TNT), pentaerythritol tetranitrate (PETN), and RDX (1,3,5-Trinitro-1,3,5-triazinane) explosives (Wu et al., 2013). In DAPCI (Jjunju et al., 2015) corona discharge is used to produce ions from nitrogen sheath gas and a mixture of methanol/water (1:1) (Figure 9.4). Reagent ions thus produced react with desorbed analyte molecules forming protonated or deprotonated molecules, depending on the mode of ionization (Williams et al., 2006).

In the DAPCI process the analyte molecule on the sample surface is ionized/desorbed using an atmospheric corona discharge to produce species, e.g., electrons, protons, metastable atoms, hydronium ions, protonated solvent ions, etc. These are further used for mass spectrometric analysis (Williams et al., 2006).

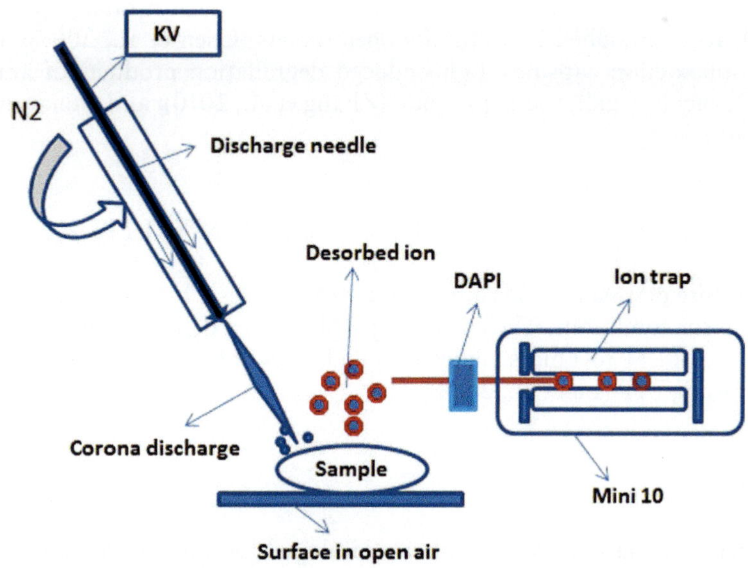

FIGURE 9.4 Schematic diagram of desorption atmospheric pressure chemical ionization (DAPCI) for direct analysis of the sample using a miniature mass spectrometer.

9.5.2 Advantages

1. DAPCI provides increased sensitivity toward compounds of moderate polarity (Wu et al., 2013).
2. The DAPCI process utilizes ambient air (without other gas assistance) as a chemical reagent for the analysis of volatile and semi-volatile compounds and smaller molecules (Williams et al., 2006).
3. For the improvement of ionization efficacy of the analyte in the DAPCI method, solvents are extensively utilized as chemical reagents, especially in case of large molecular/low volatile compounds (Yang and Deng, 2016).
4. It broadens the range of ambient MS analytes to include neutral and nonpolar compounds (Yang and Deng, 2016).
5. Reagent ion is created by utilizing atmospheric plasma directed toward the sample surface for chemical desorption (Yang and Deng, 2016).

9.5.3 Examples

1. DAPCI-MS successfully detected and identified active pharmaceutical ingredients such as aspirin, caffeine, paracetamol in Anadin Extra tablets, ibuprofen in ibuprofen gel, and cinchocaine hydrochloride and hydrocortisone in proctosedyl ointment, the active pharmaceutical ingredients of codeine and paracetamol in Solpadeine Max tablets, metoclopramide in metoclopramide tablets, and cinchocaine hydrochloride and hydrocortisone in proctosedyl ointment (Williams et al., 2006) as well as more than 20 ginsenosides in red, white, and unboiled ginsengs (Yue et al., 2012) using methanol and water (v/v = 1:1) as a chemical reagent.
2. Fingerprints for Liuwei Dihuang Teapill (Zhang et al., 2010) and *Fructus schisandrae* (Pi et al., 2011) were established by DAPCI.
3. DAPCI-MS predominately detected loratidine in Claritin tablets (Chen et al., 2007), acetaminophen in acetaminophen tablets (Chen et al., 2007), amoxicillin in amoxicillin capsules, light-induced degradation products of amoxicillin capsule aucubin, gallic acid, paeonol, (Zhang et al., 2010), and monoterpene and sesquiterpenes.

9.6 LOW-TEMPERATURE PLASMA (LTP)

Low-temperature plasma (LTP) is a technique that ionizes a broad range of organic molecules at ambient conditions. The chemical profile from a delicate object can be detected by the coupling of LTP to a mass analyzer. A wide range of complex matrices samples can be analyzed using LTP (Lu et al., 2018).

9.6.1 Theory

A low-temperature plasma probe has been developed for ambient desorption ionization (Ding et al., 2018). Dielectric barrier discharge is induced by an AC electric field by using a specially designed electrode configuration. The desorption and ionization of surface molecules are achieved in the ambient environment by extracting low-temperature

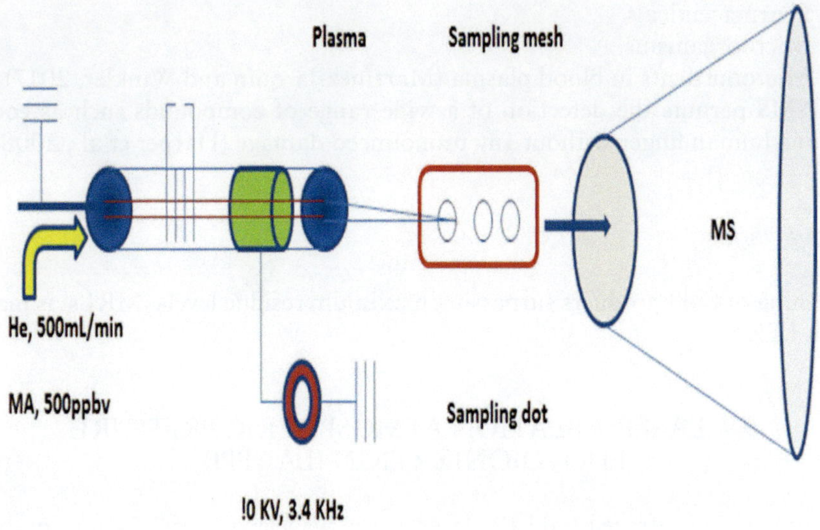

FIGURE 9.5 Schematic diagram of the reactive LTP-MS setup.

plasma from the probe interacting directly with the sample being analyzed (Figure 9.5). Thus, the experiments performed do not cause any damage to the sample or underlying substrate. For instance, biological analysis on skin surfaces can be performed without electrical shock or perceptible heating. A variety of chemical compounds may produce positive or negative ions using discharge gases such as He, Ar, N_2, or ambient air. These chemical compounds are used in the pure state and as mixtures in the gaseous or condensed phases and in solution. Limited fragmentation occurs, but this is greater in the cases of the molecular discharge gases compared to the atomic discharge gases. The characteristic mass spectra and tandem mass spectra of samples containing RDX and TNT from a poly(tetrafluoroethylene) (PTFE) surface with limits of detection as low as 5 pg demonstrated the effectiveness of LTP.

9.6.2 Applications

1. Compounds from olive oil such as free fatty acids, phenolics, and volatiles from complex, vegetable oils were evaluated by LTP-MS analysis (Martínez-Jarquín and Winkler, 2017).
2. The identification and quantification of melamine, an illicit food additive, and other potentially dangerous products from complex food and medical samples, such as whole milk, milk powder, fish paste, and urine, is done with ample sensitivity and selectivity (Huang et al., 2009 and Huang et al., 2010).
3. LTP ionization made possible the fast implementation of ambient ionization mass spectroscopy food analysis screening (Martínez-Jarquín and Winkler, 2017).
4. Detection of the following compounds is possible:
   - Explosives and warfare agents
   - Drugs of abuse
   - Coffee products
   - Crude oil

- Pharmaceuticals
- Microorganisms
- Micronutrients in blood plasma (Martínez-Jarquín and Winkler, 2017)

5. LTP-MS permits the detection of a wide range of compounds such as cocaine from a human finger without any pronounced damage (Harper et al., 2008).

9.6.3 Advantages

Fast screening of food products surpassing maximum residue levels (MRLs) is made possible with LTP.

9.7 LASER ABLATION ATMOSPHERIC PRESSURE PHOTOIONIZATION (LAAPPI)

LAAPPI uses a mid-infrared (mid-IR) laser at the wavelength of 2.94 μm to ablate the sample via its endogenous water excitation (Vogel and Venugopalan, 2003). Afterwards, a hot jet of solvent vapor is used to desolvate the ejected sample material. A krypton discharge lamp producing 10.0 and 10.6 eV photons is used to achieve photoionization of the solvent by irradiating the combined sample and solvent plume. Subsequently, gas-phase ion–molecule reactions produce the sample molecules (Figure 9.6). LAAPPI is suited for low-polarity and nonpolar compounds (Vaikkinen et al., 2014) because of its ionization mechanism (Vaikkinen et al., 2012 and Kauppila et al., 2012).

LAAPPI combines laser desorption/ablation and photoionization, and it utilizes a wavelength of 2.94 μm of the mid-infrared laser to ablate the water-rich solid samples or aqueous samples. The heated solvent vapor then desolvates the ejected sample materials. Afterward, photoionization is done by irradiating the photons generated by a krypton discharge lamp (Vaikkinen et al., 2012). Both DAPPI-MS and LAAPPI-MS are extensively applied for food products and herbal medicines analysis. Also, both DAPPI and LAAPPI possess the ability of surface imaging. DAPPI-MS successfully imaged carnosic

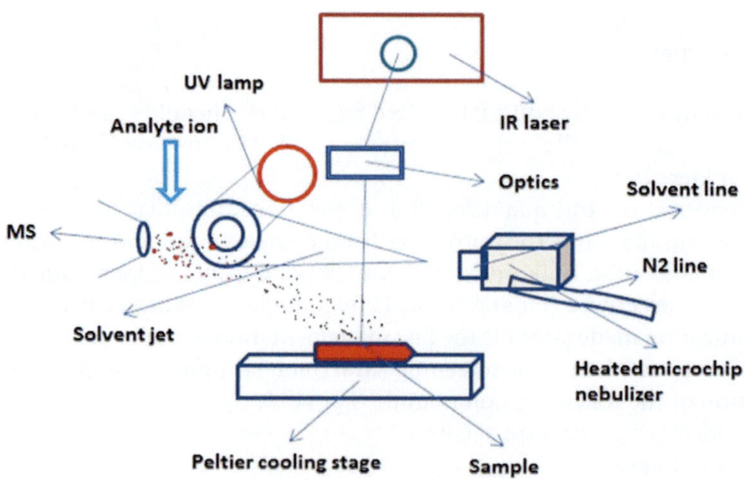

FIGURE 9.6 Schematic diagram of the LAAPPI ion source.

acid and its derivates as well as tocopherol in *Salvia* (sage) leaf, with a maximum resolution of ~1 mm (Pol et al., 2009), whereas LAAPPI-MS surface-imaged more polar terpene derivatives such as mono-sesquiterpenes and sesquiterpenes in sage leaves, with roughly 400 μm of spatial resolution (Yang and Deng, 2016).

9.7.1 Advantages

1. Electrospray-based techniques cannot ionize non-polar compound such as tricaprylin and cholesterol effectively. This obstacle has been overcome by LAAPPI, which ionizes nonpolar compounds effectively.
2. LAAPPI performs ambient analysis of nonpolar and neutral compounds, such as cholesterol and triglycerides.
3. It is able to show molecular weight-specific ions and structure specific fragments for low molecular weight (<500 amu) compounds.
4. Approximate sensitivity exists between LAAPPI and other ambient ionization methods, e.g., LAESI.
5. LAAPPI is capable of producing spectra even from a small area. For instance, a biological sample such as plant leaf and rat brain tissue section of approximately 0.08 mm^2 spots could be analyzed to produce spectra (Vaikkinen et al., 2012).

9.7.2 Applications

1. To characterize solid and liquid samples such as dried spots, biological fluids, questionable documents, counterfeit banknotes, plant and animal tissues, drug tablets, petroleum products, and thin-layer chromatography plates (Cheng et al., 2017).
2. LAAPPI can be applied to MSI of plant leaf tissues (Vaikkinen et al., 2014).
3. LAAPPI enabled the analysis of typical hydrocarbon phytochemicals, such as mono- and sesquiterpenes as well as more polar terpene derivatives, in sage leaves (Vaikkinen et al., 2014).
4. The study confirmed that LAAPPI can be used to explore the spatial distribution of nonpolar plant compounds typically analyzed by gas chromatography (GC) and liquid chromatography (LC)/MS (Vaikkinen et al., 2014).

9.8 CONCLUSIONS AND FUTURE PERSPECTIVES

Nowadays it is widely accepted that herbal products and food products are required for greater health. The safety, efficacy, and quality of food products are the main concerns for health due to the possible presence of impurities and degraded products. The detection of active pharmaceutical ingredients and impurities in food products and herbal products mainly depend upon reliable analytical methods for their quality assessment and control. As stated in this chapter, there are various ambient ionization techniques that have been used for the analysis of pharmaceutical products and herbal medicines. Several papers report the analysis by these techniques. Applications of ambient MS to achieve a systematic and comprehensive quality assessment have been described. Although there are several references reported, still great efforts are required for the improvement of the analysis by ambient MS. The use of high-transmission ion mobility mass spectrometry will further enhance the analytical capabilities by increasing the chemical coverage and obtaining more chemical information. The introduction of simple microextraction techniques to couple with ambient MS will certainly decrease the matrix effect and ion suppression that might occur in the ambient ionization process (Deng et al., 2014a; Deng et al., 2014b and Deng et al., 2015). The application of quantitation by ambient MS is still a quite challenging task, and further improvements are needed to get the desired result. Further increase in the spatial resolution of ambient MS imaging should continue to receive attention from the analytical chemists in the field. It is expected that more applications and further developments will be found in the near future.

REFERENCES

Caprioli, R. M. 2019. Imaging mass spectrometry: A perspective. *Journal of Biomolecular Techniques* 30 (1): 7–11.

Chen, H., Zheng, J., Zhang, X., Luo, M., Wang, Z., Qiao, X. 2007. Surface desorption atmospheric pressure chemical ionization mass spectrometry for direct ambient sample analysis without toxic chemical contamination. *Journal of Mass Spectrometry* 42: 1045–1056.

Cheng, S., Shiea, C., Huang, Y.-L., Wang, C.-H., Cho, Y.-T., Shiea, J. 2017. Laser-based ambient mass spectrometry. *Analytical Methods* 9: 4924–4935.

Deng, J., Yang, Y., Wang, X., Luan, T. 2014a. Strategies for coupling solid-phase microextraction with mass spectrometry. *Trends in Analytical Chemistry* 55: 55–67.

Deng, J., Yang, Y., Fang, L., Lin, L., Zhou, H., Luan, T. 2014b. Coupling solid-phase microextraction with ambient mass spectrometry using surface coated wooden-tip probe for rapid analysis of ultra traceperfluorinated compounds in complex samples. *Analytical Chemistry* 86: 11159–11166.

Deng, J., Yang, Y., Xu, M., Wang, X., Lin, L., Yao, Z. P., & Luan, T. 2015. Surface-coated probe nanoelectrospray ionization mass spectrometry for analysis of target compounds in individual small organisms. *Analytical Chemistry* 87: 9923–9930.

Ding, X., Garikapati, V., Spengler, B., Heiles, S. 2018. Analysis of ketone-based neurosteroids by reactive low temperature plasma mass spectrometry. *Rapid Communication in Mass Spectrometry* 32: 1439–1450.

Falco, L. F. G., Melo, C. F. O. R., de Oliveira, D. N., Guerreiro, T. M., Catharino, R. R. 2019. An LDI-MSI approach for targeted and untargeted differentiation and assessment of pharmaceutical formulations. *Talanta* 197: 92–97.

Haapala, M., Luosujarvi, L., Saarela, V., Kotiaho, T., Ketola, R. A., Franssila, S., Kostiainen, R. 2007. Microchip for combining gas chromatography or capillary liquid chromatography with atmospheric pressure photoionization-mass spectrometry. *Analytical Chemistry* 79 (13): 4994–4999.

Harper, J. D., Charipar, N. A., Mulligan, C. C., Zhang, X., Cooks, R. G., Ouyang, Z. 2008. Low temperature plasma probe for ambient desorption ionization. *Analytical Chemistry* 80: 9097–9104.

Huang, G., Ouyang, Z., Cooks, R. G. 2009. High-throughput trace melamine analysis in complex mixtures. *Chemical Communications* 5: 556–558.

Huang, G., Xu, W., Visbal-Onufrak, M. A., Ouyang, Z., Cooks, R. G. 2010. Direct analysis of melamine in complex matrices using a handheld mass spectrometer. *The Analyst* 135: 705–711.

Huberty, M. C., Vath, J. E., Yu, W., Martin, S. A. 1993. Site-specific carbohydrate identification in recombinant proteins using MALD-TOF MS. *Analytical Chemistry* 65 (20): 2791–2800.

Jjunju, F. P. M., Maher, S., Li, A., Badu-Tawiah, A. K., Taylor, S., Cooks, R. G. 2015. Analysis of polycyclic aromatic hydrocarbons using desorption atmospheric pressure chemical ionization coupled to a portable mass spectrometer. *Journal of American Society for Mass Spectrometry* 26: 271–280.

Kauppila, J. T., Kuuranne, T. E. C., Meurer, E. C., Eberlin, M. N., Kotiaho, T. R., Kostiainen, R. 2002. Atmospheric pressure photoionization mass spectrometry. Ionization mechanism and the effect of solvent on the ionization of naphthalenes. *Analytical Chemistry* 74: 5470–5479.

Kulkarni, P., Wilschut, R. A., Verhoeven, K. J. F., Van der Putten, W. H., Garbeva, P. 2018. LAESI mass spectrometry imaging as a tool to differentiate the root metabolome of native and range-expanding plant species. *Planta* 248: 1515–1523.

Lu, H., Zhang, H., Chingin, K., Xiong, J., Fang, H. 2018. Ambient mass spectrometry for food science and industry. *Trends in Analytical Chemistry.* doi:10.1016/j.trac.2018.07.017

Luosujarvi, L., Arvola, V., Haapala, M., Saarela, V., Franssila, S., Kotiaho, T., Kostiainen, R., Kauppila, T. J. 2008. Desorption and ionization mechanisms in desorption atmospheric pressure photoionization. *Analytical Chemistry* 80: 7460–7466.

Mandal, A., Singha, M., Addy, P. S., Basak, A. 2019. Laser desorption ionization mass spectrometry: Recent progress in matrix-free and label-assisted techniques. *Mass Spectrometry Reviews* 38: 3–21.

Martínez-Jarquín, S. R., Winkler, R. 2017. Low-temperature plasma (LTP) jets for mass spectrometry (MS): Ion processes, instrumental set-ups, and application examples. *Trends in Analytical Chemistry.* doi: 10.1016/j.trac.2017.01.013.

Mugo, S. M., Bottaro, C. S. 2007. Rapid on-plate and one-pot derivatization of carbonyl compounds for enhanced detection by reactive matrix LDI-TOF MS using the tailor-made reactive matrix,4-dimethylamino-6-(4-methoxy-1-naphthyl)-1,3,5-triazine-2-hydrazine (DMNTH). *Journal of Mass Spectrometry* 42: 206–217.

Nemes, P., Barton, A. A., Li, Y., Vertes, A. 2008. Ambient molecular imaging and depth profiling of live tissue by infrared laser ablation electrospray ionization mass spectrometry. *Analytical Chemistry* 80: 4575–4582.

Nemes, P., Vertes, A. 2007. Laser ablation electrospray ionization for atmospheric pressure, *in vivo*, and imaging mass spectrometry. *Analytical Chemistry* 79(21): 8098–8106.

Nemes, P., Woods, A. S., Vertes, A. 2010. Simultaneous imaging of small metabolites and lipids in rat brain tissues at atmospheric pressure by laser ablation electrospray ionization mass spectrometry. *Analytical Chemistry* 82: 982–988.

Pi, Z., Yue, H., Ma, L., Ding, L., Liu, Z., Liu, S. 2011. Differentiation of various kinds of *Fructus schisandrae* by surface desorption atmospheric pressure chemical ionization mass spectrometry combined with principal component analysis. *Analytical Chimica Acta* 706: 285–290.

Picca, R. A., Calvano, C. D., Cioffi, N., Palmisano, F. 2017. Mechanisms of nanophase-induced desorption in LDI-MS. A short review. *Nanomaterials* 7: 75–95.

Pogam, P. L., Schinkovitz, A., Legouin, B., Lamer, A.-C. L., Boustie, J., Richomme, P. 2015. Matrix-free UV-laser desorption ionization mass spectrometry as a versatile approach for accelerating dereplication studies on lichens. *Analytical Chemistry* 87: 10421–10428.

Pol, J., Vidova, V., Kruppa, G., Kobliha, V., Novak, P., Lemr, K., Kotiaho, T., Kostiainen, R., Havlíček, V., Volny, M. 2009. Automated ambient desorption-ionization platform for surface imaging integrated with a commercial Fourier transform ion cyclotron resonance mass spectrometer. *Analytical Chemistry* 81: 8479–8487.

Schrepp, W., Pasch, H. 2003. *MALDI-TOF Mass Spectrometry of Synthetic Polymers*. Springer-Verlag. ISBN 978-3-540-44259-2.

Seng, P., Drancourt, M., Gouriet, F., La Scola, B., Fournier, P. E., Rolain, J. M., Raoult, D. 2009. Ongoing revolution in bacteriology: Routine identification of bacteria by matrix-assisted laser desorption ionization time-of-flight mass spectrometry. *Clinical Infectious Diseases* 49 (4): 552–553.

Shrestha, B., Patt, J. M., Vertes, A. 2011. In situ cell-by-cell imaging and analysis of small cell populations by mass spectrometry. *Analytical Chemistry* 83: 2947–2955.

Vaikkinen, A., Shrestha, B., Kauppila, T. J., Vertes, A., Kostiainen, R. 2012. Infrared laser ablation atmospheric pressure photoionization mass spectrometry. *Analytical Chemistry* 84: 1630–1636.

Vaikkinen, A., Shrestha, B., Koivisto, B., Kostiainen, R., Vertes, A., Kauppila, T. J. 2014. Laser ablation atmospheric pressure photoionization mass spectrometry imaging of phytochemicals from sage leaves. *Rapid Communication in Mass Spectrometry* 28: 2490–2496.

Vogel, A., Venugopalan, V. 2003. Mechanisms of pulsed laser ablation of biological tissues. *Chemical Review* 103: 577–644.

Williams, J. P., Patel, V. J., Holland, R., Scrivens, J. H. 2006. The use of recently described ionization techniques for the rapid analysis of some common drugs and samples of biological origin. *Rapid Communication in Mass Spectrometry* 20: 1447–1456.

Wu, C., Dill, A. L., Eberlin, L. S., Cooks, R. G., Ifa, D. R. 2013. Mass spectrometry imaging under ambient conditions. *Mass Spectrometry Review* 32 (3): 218–243.

Yang, Y., Deng, J. 2016. Analysis of pharmaceutical products and herbal medicines using ambient mass spectrometry. *Trends in Analytical Chemistry*. doi:10.1016/j.trac.2016.04.011.

Yue, H., Ma, L., Pi, Z., Chen, H., Wang, Y., Hu, B., Liu, S. 2012. Fast screening of authentic ginseng products by surface desorption atmospheric pressure chemical ionization mass spectrometry. *Planta Medicine* 79: 169–174.

Zhang, X., Jia, B., Huang, K., Hu, B., Chen, R., Chen, H. 2010. Tracing origins of complex pharmaceutical preparations using surface desorption atmospheric pressure chemical ionization mass spectrometry. *Analytical Chemistry* 82: 8060–8070.

Zhong, N., Cui, Y., Zhou, X., Li, T., Han, J. 2014. Identification of prohibitin 1 as a potential prognostic biomarker in human pancreatic carcinoma using modified aqueous two-phase partition system combined with 2D-MALDI-TOF-TOF-MS/MS. *Tumour Biology* 36 (2): 1221–1231.

SECTION 3

Ambient Mass Spectrometry Techniques

CHAPTER 10

Introduction to Ambient Mass Spectrometry Techniques

Raquel Sero, Maria Teresa Galceran, Encarnacion Moyano, and Leo M.L. Nollet

CONTENTS

10.1 INTRODUCTION

Cooks and co-workers introduced the term "ambient ionization" in mass spectrometry (MS) in 2004 (Takáts et al. 2004). This term grouped those emerging desorption/ionization techniques that operated at atmospheric pressure and were able to subject an object for its immediate mass spectral analysis by handling it in front of the interface without the need for sample preparation or sample pre-treatment. However, nowadays it would be more accurate to say that these methods frequently do not require sample preparation other than the sample processing that takes place during the analysis (Venter et al. 2014; Monge et al. 2013). To classify an ionization source in this new group of ambient MS techniques, it should meet the following requirements:

- The ability to perform ionization of compounds in the open air on objects of unusual shape or size, not typically amenable to direct mass spectrometry analysis.
- The capability to perform direct surface analysis, avoiding time-consuming sample preparation steps typically required in MS-based chemical analysis of solid samples.
- Easily swappable in any mass spectrometer equipped with an atmospheric pressure interface.
- The capacity to generate ions without significant in-source fragmentation, which would simplify the mass spectral data interpretation and the compound identification when analyzing complex samples.

The two first introduced ambient mass spectrometry (ambient MS) techniques were desorption electrospray ionization (DESI) (Takats et al. 2004) and direct analysis in real time (DART) (Cody, Laramee, and Durst 2005a), which boosted the development of a growing number of new ambient MS methods and a large number of acronyms. Table 10.1 summarizes the ambient MS techniques mostly used in food and environmental analysis. A number of excellent books and book chapters (Domin and Cody 2015; Gross 2017; Moyano and Galceran 2015; Kauppila and Vaikkinen 2014; Sero, Nunez, and Moyano 2016), review articles (Ding and Duan 2015; Monge et al. 2013; Espy et al. 2014; Cooks and Mueller 2013; Badu-Tawiah et al. 2013; Cody 2013; Javanshad and Venter 2017; Snyder et al. 2016; Klampfl and Himmelsbach 2015; Kauppila and Kostiainen 2017; Venter et al. 2014) and tutorials (Huang et al. 2011; Bodzon-Kulakowska et al. 2014), that have already been published under the topic of "ambient ionization", can be recommended for a further in-depth study of ambient MS techniques fundamentals and state-of-the-art. The ambient MS techniques developed until now combine different desorption methods with different ionization techniques, thus making their classification very challenging. Some authors have classified ambient MS techniques based only on the ionization mechanism (spray or jet ionization, electric discharge ionization and gas-, heat- or laser assisted desorption/ionization), although more thorough classifications have been made based on the intrinsic extraction/desorption/ionization mechanisms involved in these techniques (Black, Chevallier, and Elliott 2016). However, these divisions are sometimes debatable, especially when several of these mechanisms concur in the same technique.

Ambient MS techniques can offer advantageous characteristics to analytical laboratories working in the field of food and environmental analysis (Kauppila and Vaikkinen 2014; Chen et al. 2017; Nielen et al. 2011; Black, Chevallier, and Elliott 2016; Luo et al. 2017; Porcari et al. 2016). For instance, real-time and *in-situ* analysis, low sample

TABLE 10.1 Acronyms and Primary References for Ambient MS Techniques Used in Environmental and Food Analysis

Abbreviation	Name	First Reference
DAPCI	Desorption atmospheric pressure chemical ionization	(Song and Cooks 2006)
DAPPI	Desorption atmospheric pressure photoionization	(Haapala et al. 2007)
DART	Direct analysis in real time	(R. B. Cody, Laramee, and Durst 2005)
DBDI	Dielectric barrier discharge ionization	(Na, Zhao, et al. 2007)
DESI	Desorption electrospray ionization	(Takáts et al. 2004)
DICE	Desorption ionization by charge exchange	(Chan et al. 2010)
EASI	Easy ambient sonic spray ionization	(Haddad, Sparrapan, and Eberlin 2006)
EESI	Extractive electrospray	(H. Chen, Venter, and Cooks 2006)
ELDI	Electrospray-assisted laser desorption ionization	(Shiea et al. 2005)
FAPA	Flowing atmospheric pressure afterglow	(Andrade et al. 2008)
LAESI	Laser ablation electrospray ionization	(Nemes and Vertes 2007)
Leaf spray	Leaf spray	(J. Liu et al. 2011)
LESA	Liquid extraction surface analysis	(Vilmos Kertesz and Van Berkel 2010)
LMJ-SSP	Liquid micro junction-surface sampling probe	(Berkel, Sanchez, and Quirke 2002)
LTP	Low temperature plasma	(Cotte-Rodríguez and Cooks 2006)
MALDESI	Matrix-assisted laser desorption electrospray ionization	(Sampson, Hawkridge, and Muddiman 2006)
Nano-DESI	Nano-desorption electrospray ionization	(Roach, Laskin, and Laskin 2010b)
ND-EESI	Neutral desorption extractive electrospray	(H. Chen, Wortmann, and Zenobi 2007)
PADI	Plasma-assisted desorption ionization	(Ratcliffe et al. 2007)
PESI	Probe electrospray ionization	(Hiraoka et al. 2008)
PS	Paper spray	(H. Wang et al. 2010)
REIMS	Rapid evaporative ionization mass spectrometry	(Schäfer et al. 2009)

requirements with little sample invasion, fast and high-throughput analysis, minimal or no sample prior preparation, small or no use of organic solvents and relatively low matrix effects are some of the ambient MS characteristics that can be attractive for food and environmental applications. These features allow some requirements to be faced such as workload, turnaround time and cost per sample frequently demanded by modern analytical laboratories. Table 10.2 summarizes some of the most recent ambient MS applications in food and environmental analysis. As can be seen, most of these applications used DESI and DART, probably because they were the first techniques developed and they have been more widely studied than others introduced later. Furthermore, the availability of commercial devices for both DESI and DART, which can be coupled to many mass spectrometer brands, may also result in the increase of their use. The simplicity of ambient

TABLE 10.2 Applications of Ambient MS Methods in Environmental and Food Analysis

		Food Analysis		
Ambient MS Technique	**Mass Analyzer**	**Sample**	**Scope of the Analysis**	**References**
DESI	IT, Q-LIT, QqQ, Miniature MS	Fruits, vegetables, cereals, olive oil, fish, meat, wine, coffee	Analysis of pesticides and lipids, food authentication and food profiling	(Cajka, Riddellova, Zomer, et al. 2011; Gerbig and Takáts 2010; Garcia-Reyes, Jackson, et al. 2009; Schurek et al. 2008; Mainero-Rocca et al. 2017; Berchtold et al. 2013; Hartmanova et al. 2010; Joyce et al. 2013; Montowska et al. 2014; J. Zhou et al. 2008; B. Li et al. 2011)
EASI	Q, Q-TOF, FT-ICR	Meat and vegetable oils	Analysis of lipids, authentication and lipid profiling	(Porcari et al. 2016)
PS	Q, IT, QqQ	Fruits, vegetables, coffee, meat, milk, olive oil and sport drinks	Analysis of pesticides, drugs and foodstuffs contaminants, authentication, metabolomics, food profiling	(Evard et al. 2015; Garrett, Rezende, and Ifa 2013; Guo et al. 2017; Y. Zhang and Lee 2013)
Leaf spray	Q-Orbitrap	Plants	Discrimination analysis	(Pereira et al. 2017)
LESA	IT	Meat	Proteomics and peptidomic analysis	(Montowska et al. 2015)
DART	QqQ, TOF, Orbitrap, QTOF, LIT-Orbitrap	Fruits, vegetables, cereals, sweets, fish, meat and beverages	Analysis of pesticides, xenobiotics, mycotoxins, phytohormones, melamine and caffeine, authentication, metabolomics, screening and characterization	(Avula et al. 2015; Kalachova et al. 2011; Crawford and Musselman 2012; Farré, Picó, and Barceló 2013; Novotná et al. 2012; Hrbek et al. 2014; J. Yang et al. 2015; Danhelova et al. 2012; Lojza et al. 2012; Rajchl et al. 2013; Luo et al. 2017)
DAPPI	QTOF, IT	Fruits	Analysis of pesticides, screening of contaminants	(Räsänen et al. 2014; Suni et al. 2012; Luosujärvi et al. 2010)
DAPCI	IT	Fruits	Screening and discrimination analysis	(S. Yang et al. 2009)
LTP	IT, Miniature MS	Fruits, vegetables, wine, coffee, milk and fish	Analysis of pesticides, melamine and volatile compounds and authentication analysis	(Garcia-Reyes, Mazzoti, et al. 2009; Gerbig et al. 2017)

(*Continued*)

TABLE 10.2 (CONTINUED) Applications of Ambient MS Methods in Environmental and Food Analysis

Ambient MS Technique	Mass Analyzer	Sample	Scope of the Analysis	References
Food Analysis				
EESI	Q-LIT, Q-TOF	Fruits, vegetables, olive oil, beverages and cheese	Analysis of lead complexes, screening and authentication	(H. Zhang et al. 2017; Bai et al. 2012; Sun et al. 2015)
REIMS	Q-TOF	Meat	Authentication/classification	(Verplanken et al. 2017)
ELDI	IT	Fungus	Determination of chemical surface composition	(M. Z. Huang et al. 2012)
Environmental Analysis				
DESI, nano-DESI	IT, Orbitrap, LIT-Orbitrap	Consumer goods, clams, air, plants, soil, aerosols and water particles	Analysis of pesticides, phthalates, toxins and PAHs	(Ewing et al. 2015; Gerbig et al. 2015; Cain et al. 2014; Boone et al. 2015; Mattarozzi et al. 2016)
PS	IT, QqQ, Orbitrap	Water	Analysis of amines, malachite green, crystal violet and their metabolites	(Jjunju et al. 2016; Fang et al. 2016)
LAESI	IT, Q-Orbitrap	Mussels	Analysis of domoic acid	(Beach et al. 2016)
DART	TOF	Lake water and aerosols	Analysis of pesticides and PAHs	(X. Wang et al. 2014; S. Zhou, Forbes, and Abbatt 2015)
DAPCI	IT	Soil	Analysis of plasticizers	(N. N. Wang et al. 2014)
DBDI	IT, Q-LIT, LIT-Orbitrap	Spiked water	Analysis of pesticides and drugs	(Mirabelli, Wolf, and Zenobi 2016)
EESI	IT	Water	Analysis of malachite green and tetrabromobisphenol A	(Tian et al. 2014; Fang et al. 2016)

MS methods allows the easy modification of the already existing ones to adapt them to both the sample type and the analytical problem. For this reason, many prototypes, home-made new approaches, are continuingly being introduced, and most of them have been explored for their application to target analysis of organic contaminants in food and environmental samples, food fraud, food authentication and screening of emerging pollutants, among others. Although most of the characteristic aspects of ambient MS techniques can be very attractive and "revolutionary", one should be aware of intrinsic limitations. For instance, the detection of a compound largely depends on the ionization mechanism as well as on the nature of the analyte to yield ions under the concrete ionization mechanism. Therefore, not a single ionization method, especially when used under one set of conditions, can deliver ions of all constituents in a complex sample. However, all ionizable compounds on the sample surface will contribute to the generation of complex mass spectral data. On the other hand, ambient ionization is generally a soft ionization that yields ions with low internal energy, which suffer no or little fragmentation in the atmospheric pressure region. Mono-charged and multiple-charged ions, as well as adduct ions, can be expected from those ambient MS methods shearing electrospray-like ionization mechanisms, while plasma-based techniques only will generate mono-charged ions and radical molecular ions. Under this scenario, the quality of the results would depend on the degree of information that the mass spectrometer can provide. Ambient MS techniques have been coupled to low-resolution mass analyzers (quadrupole and ion traps) mainly for the analysis of target compounds. However, the complexity of mass spectral data and the application of ambient MS techniques for the screening of complex samples require the use of MS instruments capable of acquiring data at high resolution and/or performing tandem MS experiments. High-resolution MS instruments are required to overcome interference problems, avoid overlapping isotope clusters and provide high-quality mass spectral information to identify both the chemical formula and the chemical structure. Today, highly sensitive and selective instruments, such as linear ion trap (LIT), time-of-flight (TOF), Orbitrap and hybrid instruments such as triple quadrupole (QqQ), quadrupole-TOF (Q-TOF), quadrupole-Orbitrap or LIT-Orbitrap, have been demonstrated to provide the necessary performance to design reliable methods with minimal sample preparation, thus facilitating high-throughput analysis.

This book chapter aims to provide a general picture of ambient MS techniques used for the analysis of organic compounds in environmental and food samples. Classification and a brief description of the ambient MS methods used in these fields as well as the most important and critical features of the desorption/ionization techniques are included.

10.2 AMBIENT MASS SPECTROMETRY TECHNIQUES

What really differentiates ambient MS from other direct mass spectrometry techniques is that the sample processing takes place in real time and proximal to ionization. Among the ambient MS techniques, the difference relies on the way in which sample processing is combined with well-known ionization mechanisms. For instance, electrospray (ESI), sonic spray (SSI), gas-phase ion–molecule reactions and photochemical ionization have been coupled in real time with sample-processing steps such as liquid extraction, thermal desorption and laser ablation, among others. The classification of ambient MS techniques in subclasses is not easy and differs depending on the reviews published until now, with a certain degree of overlap. In addition, it is sometimes difficult to distinguish between a new ambient MS technique and the simple rebranding of an already reported ambient ionization source. Despite the difficulties in categorizing the ambient MS techniques, the approximations of Venter et al. (Venter et al. 2014) and Monge et al. (Monge et al. 2013) seem to provide the most logical arrangement since both the sample processing and the desorption/ionization mechanisms are taken into account. Venter et al. (Venter et al. 2014) classified ambient MS techniques based on the predominant sample processing method: liquid extraction (spray desorption, liquid junction and substrate spray), thermal desorption and laser ablation. On the other hand, the classification of Monge et al. (Monge et al. 2013) is based on the intrinsic desorption/ionization mechanism of surface ambient MS techniques. They grouped as one-step techniques those where desorption works in parallel with the ionization, while they classified as two-step techniques those where desorption/ablation is followed by a secondary independent ionization step. It must be mentioned that these authors do not include paper spray (PS) and extractive electrospray (EESI) in their classifications. In this work, we propose a classification (Table 10.3) based on the number of steps necessary to desorb/ionize the analytes from the sample (Monge et al. 2013), but also considering the sample processing that takes place in each technique (Venter et al. 2014). Thus, this classification allows the inclusion of PS and EESI, which are classically considered ambient MS techniques.

10.2.1 One-Step Ambient MS Techniques

As mentioned above, in this group of techniques sample processing occurs in parallel with the ionization mechanism. Liquid-solid extraction, thermal desorption or chemical sputtering occurs in the same space and time as the ionization of compounds desorbed from the sample surface.

TABLE 10.3 Classification of Ambient MS Techniques Used in Environmental and Food

	Sample Processing		Ionization	Ambient MS Technique
One-Step	Liquid-Solid Extraction	Spray Desorption	ESI	DESI, EESI
			Sonic Spray	EASI
		Liquid Microjunction	ESI	LMJ-SSP, LESA Nano-DESI
	Thermal/Chemical Desorption		Plasma-based	DART, DAPCI, FAPA, LTP, DBDI, PADI,
			Photoionization	DAPPI
Two-Step	Liquid-Solid Extraction	Spray Desorption	ESI	ND-EESI
		Substrate Spray	ESI	PESI
		Ultrasonic Nebulization	APPI	EAPPI
	Laser Desorption/ Ablation	Laser Ablation	ESI	LAESI (ELDI)
		Matrix-Assister Laser Desorption	ESI	MALDESI
Three-Step	Solid-Liquid Extraction	Substrate Spray	ESI	PS
Others	Thermal Ablation/Chemical Ionization			REIMS

10.2.1.1 Solid-Liquid Extraction

Those one-step ambient MS techniques having liquid-solid and liquid-liquid extraction as the most important sample-processing steps, can be further divided into two subgroups depending on how the extraction product is presented to the ionization event: spray desorption or liquid microjunction. DESI (Takats et al. 2004) is the most well-known spray desorption ambient MS technique (Figure 10.1), in which the sample processing starts by generating a charged spray plume through pneumatically assisted electrospray of a solvent. This charged spray plume is directed onto the sample surface to create a microlocalized liquid layer where compounds are dissolved/extracted. Finally, after the impact of subsequently arriving primary charged droplets on the surface, the liberation of secondary droplets containing analytes is produced, thus generating analyte ions through traditional electrospray ionization mechanisms. Contrary to DESI, easy ambient sonic spray (EASI) (Haddad, Sparrapan, and Eberlin 2006) operates free of high voltage, in spite of sharing the same practical setup (nebulizing gas flow and polar solvent systems). The nebulizer gas flows at sonic speed (two to five times higher than DESI) coaxially to a solvent flow, which generates charged droplets through a statistical imbalance of charge. Although DESI is commercially available for installation in several mass spectrometers, EASI is still a homemade device. However, a DESI commercial ionization source can be adapted to perform EASI analysis with few custom readjustments. Regarding extractive electrospray ionization (EESI) (Chen, Venter, and Cooks 2006), the main difference when comparing with DESI and EASI is the use of two orthogonal sprays; the first spray contains charged droplets generated by the electrospray of a solvent, and the second spray can be either a neutral aerosol produced from a liquid sample or a gas stream containing volatile compounds (Figure 10.2). In this technique, collisions between neutral (containing analytes) and solvent charged droplets produce the extraction of analytes, and subsequently the ionization takes place through ESI mechanisms.

Regarding liquid microjunction-based techniques, a solvent is delivered through a capillary to form a semistatic liquid junction where analytes are dissolved/extracted from sample surface. The extracting liquid is transferred to a position where analytes can be ionized by well-known ionization mechanisms, typically by electrospray. Techniques

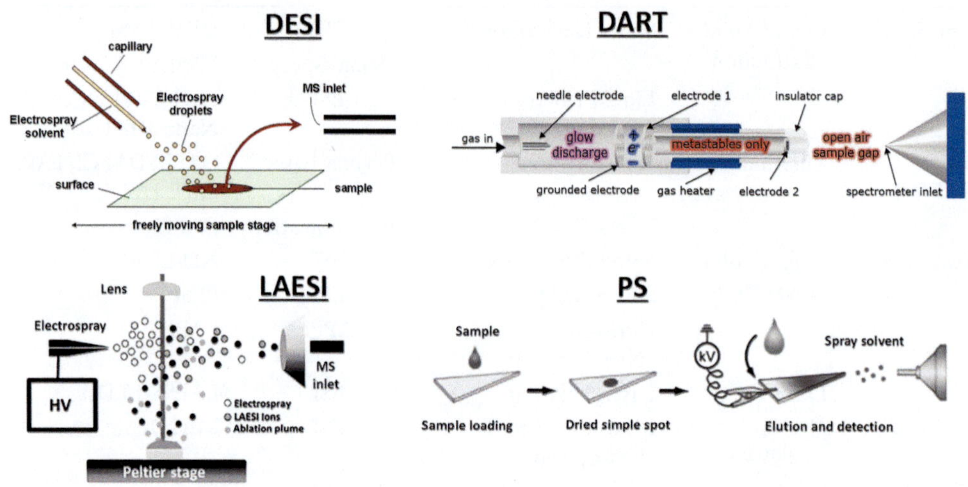

FIGURE 10.1 Schematic illustrations of some commercially available ambient MS techniques. (Reprinted with permission from Stopka et al. [2014] and Wang et al. [2010].)

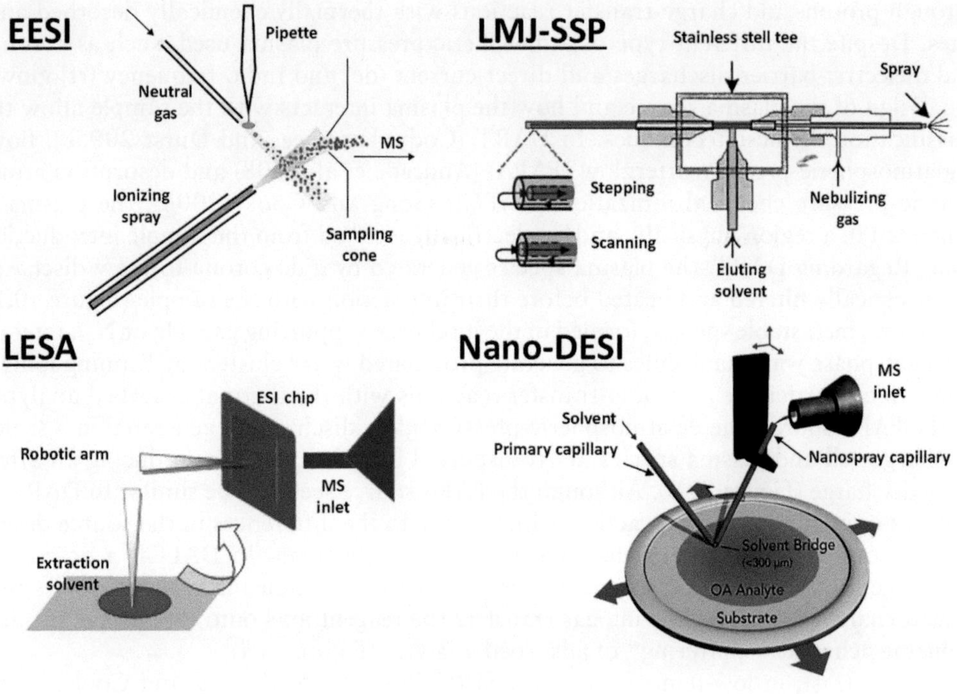

FIGURE 10.2 Schematic views of some one-step spray-based setup techniques. (Reprinted with permission from Li et al. [2011]; Berkel, Sanchez, and Quirke [2002]; Montowska et al. [2014]; and Roach, Laskin, and Laskin [2010a].)

such as liquid microjunction surface sampling probe (LMJ-SSP) (Berkel, Sanchez, and Quirke 2002), nano-DESI (Roach, Laskin, and Laskin 2010b) and liquid extraction surface analysis (LESA) (Kertesz and Van Berkel 2010) belong to this group of ambient MS methods (Figure 10.2). The LMJ-SSP, already commercialized as flowprobe™, uses two concentric tubes to supply the extraction solvent and to suction out the extracting liquid-phase with analytes, which are further ionized by ESI or atmospheric pressure chemical ionization (APCI). Nano-DESI operates more similarly to LMJ-SSP than to DESI, since the lack of nebulizing gas removes the momentum transfer of incoming droplets to the liquid film. Instead, a solvent delivered by a fused-silica capillary dissolves the compounds on the sample surface and forms a solvent bridge with a self-aspirating nanospray emitter. Regarding LESA, which is fully automated, microliquid extraction from a solid surface is combined with a nano-electrospray, thus being considered an adaptation of the infusion nano-ESI automated device (Nanomate®).

10.2.1.2 Thermal and Chemical Desorption

Thermal-assisted desorption is the most common way to remove analytes from the sample surface before gas-phase ionization through plasma-base techniques and atmospheric pressure photoionization (APPI). However, other mechanisms such as sputtering, where the sample surface is bombarded with high-energy species generated in atmospheric pressure plasmas, can also contribute to removing compounds with very low or no vapor pressure.

The plasma-based techniques falling into this group rely on an electrical discharge involving metastable and reactive charged species, which interact, directly or indirectly, through proton- and charge-transfer reactions with thermally/chemically desorbed analytes. Despite the different type of atmospheric pressure plasma used, such as coronas and dielectric-barrier discharges and direct-current (dc) and radio frequency (rf) glows, the design of the plasma source and how the plasma interacts with the sample allow the classification of these techniques. In DART (Cody, Laramee, and Durst 2005b), flowing atmospheric-pressure afterglow (FAPA) (Andrade et al. 2008) and desorption atmospheric pressure chemical ionization (DAPCI) (Song and Cooks 2006), the plasma is generated in a region physically and/or electrically isolated from the sample introduction zone. Regarding DART, the plasma species generated by a dc corona-to-glow discharge are electrically filtered and heated before their interaction with the sample (Figure 10.1). Therefore, metastable species, formed in the discharge supporting gas (He or N_2), interact with gas-phase water molecules to generate protonated water clusters by Penning ionization, which participate in proton-transfer reactions with the thermal desorbed analytes. In the FAPA source, the dc atmospheric-pressure glow discharge is generated in a sealed discharge cell and plasma species are transported into the open air producing an afterglow discharge (Figure 10.3). Although the FAPA source seems to be similar to DART, it behaves quite differently in practice, which is due to the differences in the source design and the operating conditions that will be discussed later on. In DAPCI, a dc corona discharge ionizes gas-phase solvent vapors, as in APCI, producing reagent species in a heated chamber, and the flowing gas transfers the reagent ions onto the sample surface, inducing "chemical sputtering" of adsorbed analytes (Figure 10.4).

By contrast, in low-temperature plasma (LTP) (Cotte-Rodríguez and Cooks 2006) and dielectric barrier discharge ionization (DBDI) (Na et al. 2007), the plasma is generated in the sample region and all reactive species are employed for compounds ionization (Figure 10.3). In these techniques, a "cold" (~30°C) nonequilibrium plasma is generated by a dielectric barrier discharge, sustained by an inert gas, so no damage to the surface due to heating is expected. Moreover, plasma-assisted desorption (PADI) (Ratcliffe et al. 2007) uses an rf glow discharge plasma in direct contact with the sample to desorb/ionize

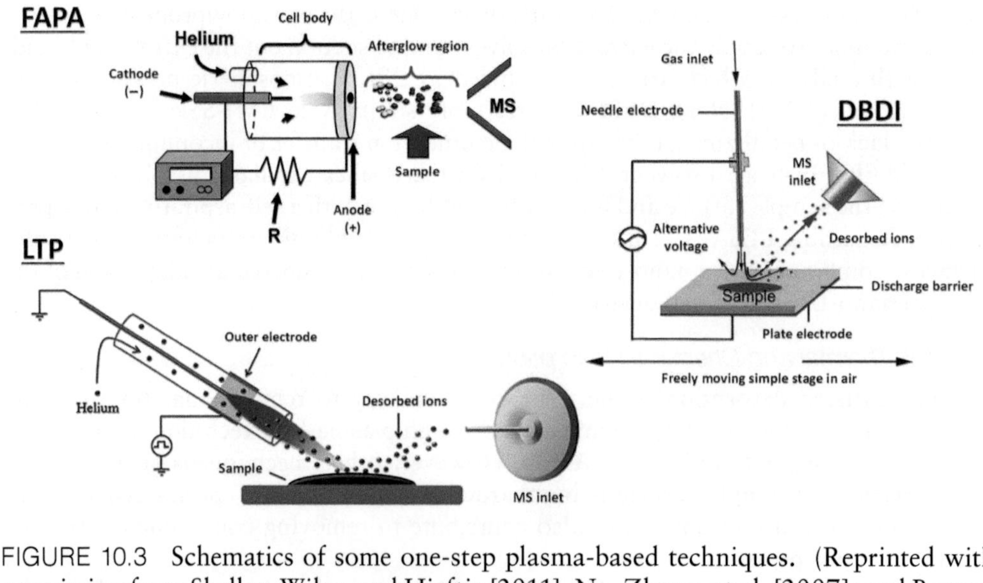

FIGURE 10.3 Schematics of some one-step plasma-based techniques. (Reprinted with permission from Shelley, Wiley, and Hieftje [2011]; Na, Zhang, et al. [2007]; and Benassi et al. [2013].)

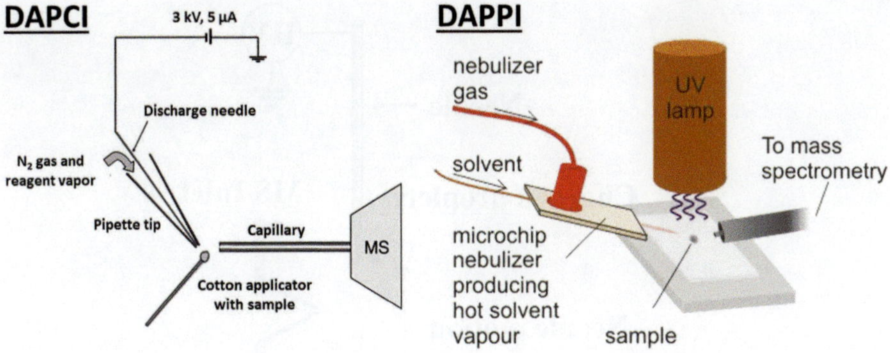

FIGURE 10.4 Schematic illustrations of DAPCI. (Reprinted with permission from Song and Cooks [2006] and DAPPI.)

target analytes. As in LTP, the plasma plume in PADI operates at room temperature, which favors the use of both techniques for the analysis of thermally labile compounds.

Finally, DAPPI (Haapala et al. 2007) (Figure 10.4) is the only non-plasma-based technique included in this group, in which analytes thermally/chemically desorbed are ionized by atmospheric pressure photoionization (APPI) mechanisms. Thus, a heated mix containing the carrying gas and solvent vapors (dopant) is directed onto the sample surface. The analytes are desorbed by the hot vapor, after which the ionization is produced by UV radiation involving photoionization, charge-transfer reactions and ion-molecule reactions with solvent/dopant species.

10.2.2 Two-Step Ambient MS Techniques

In the ambient MS techniques included in this group, the sample is nebulized and/or analytes are desorbed/extracted through thermal desorption, mechanical ablation or laser desorption/ablation in a first step that is followed by an independent secondary ionization step.

10.2.2.1 Liquid-Solid Extraction

Extraction atmospheric pressure photoionization (EAPPI) (Liu et al. 2016) is a two-step technique that uses an ultrasonic nebulizer for sample processing (extraction, nebulization and vaporization). The generated aerosol is transported by a carrier gas to the photoionization region (vacuum UV lamp), while mixing with a gaseous dopant. By contrast, in probe electrospray ionization (PESI) (Hiraoka et al. 2008) (Figure 10.5), which involves a solid-sampling electrospray probe, a needle is inserted into a sample to pick up or to coat the needle surface with sample material before positioning it close to the mass spectrometer inlet. Finally, the ionization step is performed by applying a high voltage to the needle-probe to induce an electrospray.

10.2.2.2 Laser Desorption/Ablation

This group of techniques uses an IR or UV laser to produce the desorption or ablation of analytes from a surface with and without a matrix. The sample surface is broken into small pieces by the sudden delivery of energy. The fine clusters or particles are dispersed by the expanding plume, which is subsequently merged with an electrospray plume or a

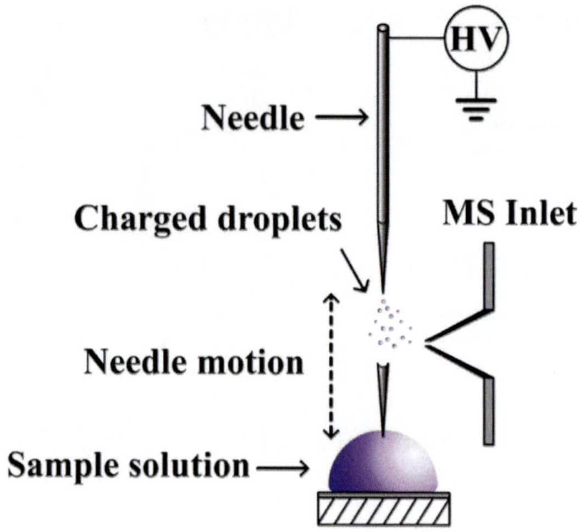

FIGURE 10.5 Schematic setup of PESI.

plasma stream depending on the ionization source used in the second step. One advantage of the sample not being in direct contact with the ionizing plume is that desorption and ionization can be optimized independently.

The first technique that coupled laser sampling to an ESI source was electrospray-assisted laser desorption ionization (ELDI) (Shiea et al. 2005). In ELDI, a pulsed nitrogen laser is used, while in laser-ablation electrospray ionization (LAESI) (Nemes and Vertes 2007) (Figure 10.1), which is commercially available, the desorption/ablation takes place applying a mid-IR laser. By contrast, matrix-assisted laser desorption electrospray ionization (MALDESI) (Sampson, Hawkridge, and Muddiman 2006) uses an IR or UV laser to excite an exogenous matrix that cocrystallizes with the analyte, while applying a voltage to the stainless-steel target plate.

10.2.3 Three-Step Ambient MS Techniques

In this classification, a three-step ambient MS group has been proposed to include paper spray (PS) as an ambient MS technique, since it incorporates two sampling/processing steps and the ionization as a third step. In paper spray (PS) (Wang et al. 2010; Liu et al. 2015), a prepared liquid sample is spotted onto a paper triangle, and a solid phase extraction and/or chromatographic separation occurs when delivering a solvent and applying a high voltage to the paper substrate (Figure 10.1). The extracting solvent transports the extracted compounds to the apex of the triangle paper, whereas the different interactions of sample components with the paper substrate could produce their spatial/temporal separation. Electrospray is the driven force that pulls the solvent to the apex of the paper triangle, but also it is responsible for compound ionization through ESI mechanisms. Leaf (Liu et al. 2011; Liu et al. 2015) is an ambient MS method sharing similar principles with paper spray. However, in this technique the plant material serves both as a substrate and as a sample. For instance, plant materials having a natural sharp shape (e.g., bean sprout) can be analyzed directly by leaf spray, while a small nick cut with a sharp tip has to be made to non-pointy samples. As in PS, the spray solvent is supplied on the pointy plant material to extract the endogenous compounds and to transport them to the tip. Finally, by applying a high voltage directly to the plant material, a spray of charged droplets is generated at the tip through ESI mechanisms.

10.2.4 Other Ambient MS Techniques

10.2.4.1 Rapid Evaporative Ionization Mass Spectrometry (REIMS)

REIMS is an ambient MS technique based on other principles for desorption or ionization that cannot be classified in any of the previous categories. Originally, this technique was developed for the real-time identification of tissues during surgical interventions (Schäfer et al. 2009), although it has recently been proposed as a new analytical approach for the direct analysis of meat samples (Balog et al. 2016; Verplanken et al. 2017). In REIMS, surgical blades are used to ablate tissues in order to produce aerosols and charged species by the heat dissipated during the electrosurgical process, being subsequently transported to the MS inlet. The ionization mechanism proposed for REIMS is quite similar to that taking place in both APCI and thermospray ionization in filament-off mode (Schäfer et al. 2009). The desorption of neutral molecules is followed by gas-phase ionization *via* proton transfer reactions with the ionized water molecules as in APCI.

10.3 PHYSICAL DESIGNS AND EXPERIMENTAL APPROACHES

As mentioned above, in ambient MS analysis, sample processing and ionization usually take place in a single platform near the inlet of the mass spectrometer. In this context, geometrical parameters, solvent and/or gas characteristics and flow rate, plasma formation approaches, surface properties, solubility and temperature are important variables that critically affect the performance of the techniques.

10.3.1 Probe Assembly Configurations

In one-step spray desorption-based techniques (Table 10.3) such as DESI, EASI and EESI, the geometrical configuration and operational parameters are important aspects to be considered (Weston 2010). For instance, when tuning a DESI source, the response depends on the incident and the collection angles, the sample spot-to-MS inlet distance, as well as on both the tip-to-surface and the MS orifice-to-surface heights (Figure 10.1). Recommended parameters for DESI typically are within the range of 4560° and 530° for incident and collection angles respectively, 23 mm for the sample spot-to-MS inlet distance and 13 mm and ~1 mm for the tip-to-surface and the MS orifice-to-surface heights, respectively (Monge et al. 2013). The liquid sample in EESI (Figure 10.2) is sprayed into an electrospray plume positioned orthogonally to the sample spray, the distance and the angle between both sprays being critical to guarantee an efficient overlapping. The geometrical configuration of liquid microjunction-based techniques is quite different and the sampling probe rests on the surface of the sample for a longer time than in spray-based ones. The differences among these techniques are due to the method of delivering and removing the extracting solvent from the sample surface. As can be seen in Figure 10.2, in LMJ-SSP the coaxial capillaries are positioned perpendicularly (100300 μm) above the sample. The outer capillary supplies the solvent to form the liquid junction and extracts compounds from the sample surface, whereas the inner capillary pulls the solution that contains the analytes to the ionization source. In contrast to the vertical arrangement of LMJ-SSP, in nano-DESI the two fused silica capillaries that form the liquid microjunction are positioned at an angle above the sample surface (Figure 10.2). The primary capillary continuously supplies the solvent to create and maintain the liquid-bridge, while the secondary capillary, an electrospray emitter, aspirates and ionizes the desorbed analytes. Regarding LESA (Figure 10.2), a robotic arm picks up a conductive disposable pipette tip and moves it above to a well containing the extracting solvent, which is aspirated by immersing the pipette tip into it. Afterwards, the pipette tip is positioned above the surface spot to form the liquid microjunction by dispensing a specific volume of solvent on the surface to extract the compounds. The latter liquid-phase is aspirated back into the tip and positioned in one of the emitters of a chip-based nano-ESI for the ionization. The main advantages of LESA are the full automatization and the use of disposable tips, which prevents contamination and carry-over.

As mentioned before, the method performance of plasma-based techniques is more affected by the physical design of the plasma source and the kind of plasma–sample interaction than by geometrical configurations (Ding and Duan 2015). Thus, in LTP and PADI the plasma is generated in the sampling zone, while in DART, FAPA and DAPCI the plasma is isolated from the sample ionization region. The DART source (Figure 10.1) consists of two chambers through which the DART gas flows. In the first region, a dc corona-to-glow discharge (15 kV) generates the plasma, which is filtered by one or several perforated plate electrodes and heated before passing through a final grid electrode to interact with compounds thermally desorbed from the sample surface. This design allows the selective removal of ionic species from the heated plasma gas (typically He); thus only electronically excited metastable atoms arrive to the outside region. In DART, the sample is placed near the source outlet with minimum disturbance of the gas flow to the MS inlet, geometries for direct desorption and for transmission being commercially available. By contrast, in DAPCI, the plasma species are transferred to the sample region without any filtration (Figure 10.4). This ambient source is built by simple modifications of the conventional APCI source, where gaseous reagent ions, generated by atmospheric corona discharge, are sprayed through a pipette tip and impact on the solid sample, thus producing the desorption and subsequent ionization of analytes. In FAPA, LTP and DBDI, a two-electrode configuration with different power supplies, electrode shape and discharge cells is used. For instance, FAPA operates in a current controlled glow discharge regime (~ 25 mA) applying a direct voltage (typically a few hundred volts) to pin-to-plate or pin-to-capillary electrodes in a sealed discharge cell that physically separates the source from the sample (Figure 10.3). Unlike DART, FAPA does not use any electrode to filter plasma species before interacting with the sample. Moreover, no external heating is needed in FAPA, since the temperature of the plasma stream is high enough (> 200°C) for an efficient desorption of the analytes. Regarding ambient MS methods based on dielectric barrier discharge (DBD), at least one dielectric layer has to be placed in between the electrodes, a high ac voltage being necessary to transport the current through the discharge gap. Typically, amplitudes of 1100 kV and frequencies of a few Hz to MHz are commonly used. The difference between DBDI and LTP lies in the geometrical configuration. In DBDI source, a glass slide is inserted between a plate electrode and a needle electrode to function as dielectric material and sample plate (Figure 10.3). When ac high voltage (510 MHz) is applied between the electrodes, a low-temperature plasma consisting of numerous transient micro-discharges is ignited and analytes placed in the surface plate are desorbed and ionized. On the other hand, LTP uses a glass tube (typically, ~4 mm inner diameter and ~6 mm outer diameter) as dielectric material to physically separate the ring outer electrode from the metal pin internal grounded electrode, which is directly in contact with the flowing gas (Figure 10.3). The generated plasma interacts with the sample, thus desorbing and ionizing analyte molecules. Since thermal desorption processes clearly play a major role in plasma-based techniques, temperature becomes a critical parameter to improve sensitivity. As commented before, FAPA does not use additional heating since the heat generated in the source is enough to desorb analytes from the sample surface. By contrast, DART uses external heating to desorb semi-volatile and low-volatile compounds. In fact, selective detection of thermal-dependent compounds can be performed in DART by increasing the temperature of the discharge gas to produce thermal fractionation of the compounds in the sample surface. PADI is another "proximal" plasma-based technique, where high-energy plasma species interact directly with the sample. Hence, an rf glow discharge is set at the end of a pin electrode, which comes in direct contact with the sample, that works as ground electrode.

Some thermal desorption techniques are not plasma-based. Among them, DAPPI has increased its popularity in recent years because it can be applied for the analysis of low-polarity compounds (Kauppila and Kostiainen 2017). In this technique, a heated microchip is used to deliver both a narrow jet of vaporized solvent (dopant) and a nebulizer gas onto the sample surface, thus achieving surface temperatures up to 350°C for the desorption (Figure 10.4). Ionization is initiated by photons emitted by a vacuum UV lamp, typically a krypton discharge lamp that emits 10 eV photons, although dopants such as toluene, acetone or anisole (typically added at 10 µL min^{-1}) are required to ionize those compounds with ionization energies below that of the photons.

For the two-step laser-based desorption/ablation techniques, the key parameters affecting their figures of merit are the characteristics of the lasers used for ablation or desorption of the analytes from the sample surface, the ionization method, the use of matrix and the geometrical configuration (Javanshad and Venter 2017). Both UV and IR lasers are used to generate the plume of particles, clusters and free molecules, which are subsequently ionized in a separated step. For instance, UV nitrogen lasers, mainly at 337 nm operating at 10 MHz with a pulse length of 4 ns and a pulse energy of 20 µJ, are used in ELDI. However, in most of the techniques, as for instance in LAESI, IR lasers typically tuned at 2940 nm with pulses of 5 ns duration at 220 Hz and pulse energy between 100 mJ and 52 mJ are used (Figure 10.1). As regards to the ionization step, the most popular is electrospray ionization, which is used in techniques such as LESI, LAESI and in MALDESI where an exogenous matrix is used to improve ionization. Although both reflection and transmission approaches are used in laser-based techniques, most of the geometrical implementations use reflective geometries with incident laser beams hitting the sample perpendicularly. However, other incident angles such as 45° with respect to the sample have also been used in some cases. In the less-used inverted geometry (transmission mode) the laser beam from the backside illuminates the sample, and the plume of desorbed material exits from the front side in the opposite direction from laser application.

10.3.2 Solvents, Gases and Sample Devices

In liquid extraction-based methods, a solvent is always employed and its selection is a key aspect to be considered. In spray desorption techniques such as DESI, a spray of solvent charged droplets is used to perform the desorption/extraction of compounds and to ionize the analytes, while in EESI and in liquid-microjunction techniques the solvent extraction takes place in line with an electrospray emitter. Solvent composition affects both the initial spray droplet size, which depends on both the surface tension and the dielectric constant of the solvent, as well as the capacity of the solubilizing sample component (Javanshad and Venter 2017). In general, organic/aqueous binary mixtures containing 50–80% of methanol or acetonitrile are used to dissolve analytes. Moreover, the solvent system has to be electrically conductive and typically, when working in positive ion mode, acetic or formic acid (0.1–0.5%) is used to increase ionization efficiency. DESI is generally indicated for polar compounds that can be easily ionized by controlling the pH of the solvent mixture. Even so, low-polar and hydrophobic compounds have also been analyzed using nonpolar solvents such as toluene (Green et al. 2010). In this case, the solvent is ionized by electrochemical oxidation at the metal spray needle interface with the subsequent ionization of analytes by charge transfer processes. As regards flow velocities, both sheath gas and liquid solvent flow rates, which are interdependent, affect the performance of the technique. A threshold velocity of gas flow is required for an efficient production of small secondary droplets, but values above the optimum generate a spray focused in a small spot, reducing the amount of desorbed material. This effect can be compensated by increasing the solvent flow rate. Typical gas pressure values (N_2) ranging from 120 to 150 psi and solvent flow rates from 2 to 10 μL min^{-1} are used to maximize the ion signal (Bodzon-Kulakowska et al. 2014). In nano-DESI, no sheath gas is needed since the flow rate for the liquid junction is governed by electrospray-induced flow generated by the nano-spray tip and additionally assisted by the vacuum in the MS inlet. In LMJ-SSP, typically suction flows of 10 μL min^{-1} are used by controlling the inner and outer capillaries diameters and gas flow. This technique can work in two operating modes: stepping and scanning. In stepping mode, a single spot of the sample is analyzed at a time, while in scanning mode the liquid microjunction is dragged across the surface obtaining one- and two-dimensional images. In this last case, a thorough control of flow rate is needed since low flow rates worsen the spatial resolution, while high flow rates and scanning speed decrease the signal intensity.

Today, the reduction of gas consumption in miniaturized plasma-based ion sources is an attractive feature for their implementation in mass spectrometer instruments for fieldwork. The discharge gases most frequently used are helium, nitrogen, argon or air at flow rates that depend on the technique, ranging from <1.5 L min^{-1} in pin-to-capillary FAPA sources to ~3 L min^{-1} for DART. Although higher flow rates can increase signal responses, they generally produce strong turbulences that worsen reproducibility (Ding and Duan 2015).

In ambient MS, the addition of specific reagents can be used to enhance selectivity and sensitivity for the detection of certain compounds, thus increasing the range of chemicals accessible by a particular method. This is typically done by transforming the analyte to a permanent charged product or to a product that is easy to ionize. The most popular reactive ionization-based technique is reactive DESI (Wu et al. 2009). In this case, a derivatizing agent is added to the spray solvent and chemical reactions occur in the short timescale of the solvent extraction/ionization/detection process (Laskin and Lanekoff 2016). Reactive DESI has been used for the detection of reaction intermediates that have lifetimes of the order of milliseconds (Girod et al. 2011). The addition of reagents to the solvent has also been used in most of the techniques based on electrospray ionization such as PS, EESI or ELDI in order to improve the ionization of compounds that are not easy to ionize (Venter et al. 2014). Similarly, in thermal desorption-based methods, reagent and solvent vapors can be added to modulate ionization by increasing differences in proton affinities or by generating other reactive species (Venter et al. 2014). For instance, *in-situ* silylation or methylation have been performed in DART by depositing the reagent onto the sampling device placed in the heated gas stream or by adding the reagent to the discharge gas through a T-junction as in LTP.

It is currently assumed that one of the main characteristics of ambient MS techniques is their ability to analyze solid samples in their native form, as for instance, the analysis of fruit peels by simply holding them with a pair of tweezers in the plasma of a DART source. However, samples are sometimes deposited onto auxiliary sampler devices, such as glass slides, fused capillaries, etc. The nature and characteristics of these holders can affect the performance of the methods. For instance, in DESI, nonconductive surfaces are typically used to prevent neutralization of the charged droplets at the surface. Moreover, surface wettability must be taken into account since it affects both spray–surface and analyte–surface interactions. Several materials such as glass, paper and polymers such as PTFE (polytetrafluoroethylene) and PMMA (polymethylmetacrylate) are used to spot liquid samples on them for the DESI analysis. In general, the best results are achieved with the non-polar PTFE surfaces, probably because the lower interaction with the analytes helps the desorption from the sample (Wiley et al. 2015). In DART, commercially available sampling devices included clamp-like tweezers or fingers to fix the sample (fabrics, disks, swabs, etc.) and holders for glass capillaries when analyzing liquids and powdered samples (Cody and Dane 2015). ELDI and LAESI use stainless-steel target plates although silver or gold plates are recommended to prevent the formation of metal-analyte adducts. Several ambient MS techniques can work in transmission mode by using transmissive/porous materials. In techniques such as DESI, DART and DAPPI, a screen mesh that supports or embeds the sample is placed in front of the DESI tip, the DART plasma or the microchip in DAAPI. Afterwards, the plasma or heated solvent passes through the screen mesh to desorb compounds and, in some techniques, to ionize analytes. In this mode, the material type affects the coating of the sample onto the strand surface and both metallic and polymeric meshes such as PEEK are employed as sampling substrates (Ding and Duan 2015). By contrast, laser ablation techniques use a transparent sample holder, generally a glass slide or a fused silica capillary. In this case, the laser is transmitted through the sample holder, and the plume of desorbed material exits in the opposite direction to where the laser is applied (Monge et al. 2013).

Regarding substrate spray techniques, samples or sample substrates having sharp tips (such as sticks, paper triangles, needles, etc.) are used. In PS, a metal clip, connected to a dc high voltage (3–5 kV) power supply, holds a triangular-shaped paper placed in front of the MS inlet (Figure 10.1). The characteristics of the paper substrate and the solvent have an important effect on the efficiency of the method. Paper substrate (filter paper, glass fiber paper, silanized paper, chromatography paper, etc.) influences the interaction between the analyte and the paper, and the triangle tip corner angle (30 to 45°) has a high impact on the generation of electrospray. Moreover, the nature of the solvent affects both the extraction efficiency and the electrospray ionization. Generally, PS applications use methanol/water mixtures, although less-polar solvents can also be employed to extract less-polar analytes (Liu et al. 2015). By contrast, PESI uses a solid needle to pick the sample, which is directly ionized by electrospray from the sharp tip of the probe (Hiraoka et al. 2008). One of the advantages of working with a solid needle, over other techniques that use capillary tubes, is that clogging is prevented. Disposable stainless-steel acupuncture needles with a tip diameter below 1 μm are generally used, although other kinds of conductive needles and even surface-coated wires for the selective detection of specific compounds have also been employed. The solid needle probe is mounted in a system that allows it to be driven up and down and to be adjusted to a position just touching the biological or liquid sample. The sample volume picked up depends on the size of the needle, the hydrophobicity of needle surface, the dipping depth and the viscosity of the sample. After the sample collection, the needle is moved to the upper position located in front of the MS inlet, and a high voltage (~2 kV) is applied to generate the electrospray plume (Figure 10.5). Although for biological samples the small amount of water content (~pL) is enough for the electrospray ionization, the use of auxiliary spray solvents has also been proposed to improve ionization. Needles have also been used for the direct spray of tissue samples in MS analysis biopsy, and this strategy could be applied to the analysis of meat/fish tissues (Monge et al. 2013). In this case, the spray solvent wets the tissue sample held in the needle tip, and a high voltage directly applied to the metal needle generates charged droplets containing compounds extracted from the tissue. Another technique that has been employed for the analysis of tissue materials is REIMS, where a high-frequency electric current is applied to surgical blades to produce the ablation of tissue material and to lead to the formation of an aerosol that contains charged species (Schäfer et al. 2009). In a typical setup, ions and aerosols created during electro-surgery are aspirated by a Venturi air jet pump, which allows the transportation of the thermal-generated ions to the mass spectrometer for analysis.

10.3.3 Ambient MS Imaging

One interesting application of ambient ionization MS techniques is mass spectrometry imaging (MSI) (Laskin and Lanekoff 2016). MSI enables one to acquire spatial information and identification of chemical compounds distributed on a sample surface. Although secondary ion mass spectrometry (SIMS) and matrix-assisted laser-desorption ionization (MALDI) have been extensively used for MSI, today ambient MS is also applied, taking advantage of its ability to analyze a surface at atmospheric pressure with any pretreatment. Techniques such as DESI, EASI, LMJ-SSP, nano-ESI, PESI, DAPPI, LTP and those based in laser desorption combined with ESI or with chemical ionization have been developed. For data acquisition, small areas (pixels) of the sample surface are scanned individually and sequentially, and an automated moving stage is used to assure the reproducibility of the scan velocity. Recorded mass spectra are converted to 2D images by using adequate image software. Important characteristics of these techniques are spatial resolution, sensitivity and ability to perform depth profiling. Lateral resolution mainly depends on the spot size of the ionizing agent although it is also limited by operating conditions. However, it must be taken into consideration that when working at high spatial resolution, a high acquisition time is required to overcome the decrease in sensitivity. In DESI, the characteristics of the sprayer (solvent, flow rate, nebulizer gas pressure and tip dimensions) are important to obtain a high-quality MS image. At standard conditions, lateral resolutions of 150–250 μm are obtained, although values as low as 40 μm can be found if the distance between the sprayer and the surface is decreased to 400 μm (Kertesz and Van Berkel 2008). Resolution of LMJ techniques is mainly determined by the size of the capillaries and the ability to control the distance between the sample and the probe. In LMJ-SSP, resolution values from 500 to 1000 μm are obtained (Laskin and Lanekoff 2016), while for nano-DESI, where two independent small-diameter capillaries are used, better resolutions (100–150 μm) are typically found, and values down to 12 μm have been obtained using pulled capillaries (Laskin et al. 2012). The size of the pipette tip is the main reason why LESA spatial resolution is worse than in other liquid extraction-based ambient MS techniques (~ 1 mm). In PESI, a good spatial resolution (60 μm) can be obtained by controlling the diameter of the needle probe, the space in between two adjacent holes and the depth of penetration of the needle. In PESI-imaging, an auxiliary solvent sprayer is typically used to re-wet the sample adhered to the needle to facilitate the electrospray process. For DART and related thermal desorption-based techniques, the ability to generate spatially resolved information is very limited and lateral resolutions are typically of 1 mm. The best results, resolutions of 200 μm, are obtained with LTP focusing the plasma onto a small area of the sample surface. One advantage of LTP is that depth profiling can be performed simply by increasing the ablation time of the plasma torch. Ablation can also be easily obtained using lasers and the coupling of laser ablation to electrospray for ionization of the neutrals ablated from the sample provide IMS techniques with the best lateral resolution and with the additional advantage of being able to perform *in-situ* depth profiling, thus providing a three-dimensional (3D) reconstruction of molecular distributions. The most popular techniques are LAESI and ELDI using nanosecond lasers, mid-IR and UV, respectively. In LAESI, a lateral resolution of 250 μm is typically obtained, although this value can be reduced to 50 μm when using an optical fiber to deliver the laser light. The ablated plume generated either in transmission or reflection mode can also be captured into a single drop or a continuous flow of solvent in a probe similar to the LMJ-SSP procedure, thus providing very good spatial resolutions (50 μm) in transmission mode (Laskin and Lanekoff 2016). The use of femtosecond mid-IR lasers allowed very good lateral resolutions down to 10 μm although focusing of the laser is needed to obtain reproducible results (Sarkar et al. 2009).

10.4 SAMPLE HANDLING

The advantages of ambient MS techniques, such as short analysis time (in seconds) and operational simplicity, made these techniques highly attractive for high-throughput applications, which quickly led to the development of new strategies to adapt the ambient MS technique to specific samples and analytical needs. Although the original objective of ambient MS techniques is to perform real-time analysis without additional sample preparation, with the exception of the sample processing that takes place during the analysis, sample manipulations are sometimes needed. However, this sample handling should be minimal, simple and fast to be compatible with the philosophy of ambient MS methods.

Before the analysis of food and environmental samples with ambient MS techniques, fast and simple manipulation procedures are often necessary when dealing with matrices such as solids, non-smooth surfaces, semisolids or viscous samples. For instance, powdered samples such as powdered milk and spices have to be solvent-extracted or solvent-agglomerated before the analysis with gas-assisted nebulization techniques such as DESI and DART. These simple strategies might prevent the carry-over contamination by blowing the sample, which could lead to false-positive results in the analysis of subsequent samples (Nielen et al. 2011; Zhang et al. 2015; Yang et al. 2009). For solid, high-viscosity or gelatinous semisolid samples (i.e., oils and jelly), a simple extraction or a sample dissolution is enough before their analysis by ambient MS techniques that are more suitable for liquid samples such as PS or EESI (Zhang, Cooks, and Ouyang 2012; Garrett, Rezende, and Ifa 2013; Huang et al. 2014; Liu et al. 2016). Moreover, trituration and homogenization of solid samples are sometimes recommended to guarantee the representativeness of the analyzed sample (Sero et al. 2015).

Another question to be considered is that the surface analysis of solid samples only reveals what is present on the sample surface, which may be a handicap for some food and environmental analytical purposes. For instance, when analyzing leaves with some ambient MS techniques, the plant cuticle, a wax protective layer on the upper epidermis of the leaf, prevents the desorption/ionization of compounds in inner layers of the leaf. For this reason, the removal of the wax layer by washing the leaf surface with nonpolar solvents (i.e., hexane or chloroform) and/or cryosectionation of plant tissues to detect analytes coming from the inner layers of leaves are frequent procedures before the ambient MS analysis. Other interesting procedures that can be used are imprinting strategies, where plant tissue are pressed (imprinted) on a substrate such as tape, TLC plates or porous Teflon, to transfer the plant compounds to the substrate for a later ambient MS analysis or imaging (Laskin and Lanekoff 2016; Janfelt 2015). Thermal imprinting has also been used for the EASI-MS lipid profiling of meats, fats and fish for food authentication (Porcari et al. 2016). In this strategy, the fatty tissue is wetted with a small amount of solvent and the surface is heated before the imprinting onto a piece of paper.

The use of a minimal sample manipulation/treatment is also justified when sample components cause severe matrix effects and difficult matrix interferences that are unavoidable *via* high-resolution mass spectrometry or tandem mass spectrometry. To overcome matrix interferences, fast separation strategies have also been applied in combination with ambient MS methods. In DART, the temperature of plasma gas can be varied in time to favor a thermal separation of compounds of different volatility (Edison et al. 2011). Furthermore, one important drawback of ambient MS methods lies in their sensitivity. The use of chromatographic substrates in ambient MS analysis has been proposed for a simple sample purification to improve limits of detection (LODs) of compounds to be identified in complex samples. TLC has been used as an off-line technique for sample extract purification before the ambient MS analysis. Other chromatographic techniques such as gel-permeation chromatography (GPC) and reversed-phase chromatography have been also used to separate components (Chen et al. 2017; Hajslova, Cajka, and Vaclavik 2011; Lu et al. 2017; Black, Chevallier, and Elliott 2016).

Most of ambient MS methods suffer from high limits of detection, which is not a problem for most applications such as food fraud and food profiling/authentication, but it is an important disadvantage when the analytical objective is the detection of analytes at trace levels in complex samples such as in food safety and in environmental analysis. For this reason, more extensive and time-consuming sample preparation protocols, typically used for more traditional analytical techniques, have been found to be necessary in some qualitative and quantitative applications. For instance, dispersive solid phase extraction procedures such as Quick, Easy, Cheap, Effective, Rugged and Safe (QuEChERS) protocols improved LODs and limits of quantitation (LOQs) when screening and quantifying trace levels of pesticides and mycotoxins in food samples by DESI, LTP and DBDI-MS (Chen et al. 2017; Kauppila and Vaikkinen 2014). By using these sample treatment strategies, the good sensitivity and reproducibility of the method demonstrated that ambient MS techniques can easily be used as an alternative for high-throughput pesticide residue analysis (Porcari et al. 2014; Mattarozzi et al. 2016; Mirabelli, Wolf, and Zenobi 2016). On the other hand, to determine low levels of analytes in liquid food samples, liquid-liquid extraction (LLE), solid-phase extraction (SPE), liquid-phase microextraction (LPME) and solid-phase microextraction (SPME) procedures have been established, thus allowing the clean-up of the extracts and the preconcentration of analytes (Cajka, Riddellova, Tomaniova, et al. 2011; Chen et al. 2017). These extraction procedures have been used for monitoring/screening studies and quantitative analysis of target compounds, taking advantage of sensitivity improvement (up to 700-fold) and matrix effect reduction. The distinctive future of SPME is that the fiber is frequently used as a probe in ambient MS analysis. In this context, three different strategies have been applied. In one of them, the SPME fiber is directly analyzed with spray-based ambient MS techniques by applying a high voltage and a solvent to the SPME probe, thus inducing an electrospray at the tip of the fiber to ionize the analytes *via* ESI mechanisms. The second strategy is based on the surface desorption/ionization of analytes by the impact of charged solvent species *via* DESI and LTP on the SPME fiber surface. Finally, the third strategy is based on desorption of the analytes through nebulization, laser desorption, laser ablation, thermal evaporation or thermal desorption from the SPME probe and a later ionization such as DART. The developments of highly selective adsorbents towards target compounds, novel coating materials for the improvement of both extraction and ionization efficiencies and new coupling strategies to enhance hyphenation properties are highly encouraging (Fang et al. 2016).

10.5 APPLICABILITY AND METHOD PERFORMANCE

Today, ambient MS techniques have been applied for a wide range of applications in food and environmental analysis, and the easiness of implementation and the use of ambient MS systems for on-site analysis by non-expert users bring a shift in MS applications for routine analysis in quality control laboratories. Furthermore, the new developments in the miniaturization of mass spectrometers, designed for on-site analysis, can be a valuable alternative to current in-lab MS analysis operations.

Food quality and safety assurance in the food supply chain are crucial for public health and world sustainability. Most of the food products available in supermarkets are susceptible or have already been exposed to some form of food fraud, which increases the concern of authorities and food industry. As can be seen in Table 10.2, the application of many ambient MS techniques to the food supply chain, such as production, processing, storage or transportation, has proven that ambient MS techniques can play a key role in this field, thus providing important information about food components (e.g., lipids, hormones), contaminants (e.g., pesticides, veterinary drugs, mycotoxins), authenticity and traceability (Vaclavik et al. 2009; Black, Chevallier, and Elliott 2016). Moreover, ambient MS techniques have also been applied for the rapid analysis of contaminants present in environmental samples such as polycyclic aromatic hydrocarbons (PAHs), volatile organic compounds (VOCs) and perfluorinated compounds (PFCs) in soil and environmental water (Chen et al. 2017; Luosujärvi et al. 2010). However, the need for detecting analytes at trace levels makes the direct analysis by ambient MS techniques less attractive, and sample manipulation strategies are usually required.

As mentioned before, two of the most powerful characteristics of ambient MS techniques are the ability for high-throughput analysis and the capacity to deliver a wealth of chemical information with unprecedented easiness. These main characteristics make these techniques very attractive for the screening of targeted and untargeted compounds, which is one of the main applications of ambient MS in food and environmental analysis. The screening of pesticides, phenolic compounds or triacylglycerol compounds in several food samples has been successfully achieved (Deng et al. 2017; Black, Chevallier, and Elliott 2016). However, one of the problems when performing the direct analysis of fruits, vegetables and plants is the high variability of the sample surface composition from unit to unit, which can be a problem regarding sample representativeness. As an alternative to surface analysis, porous cotton or polyester swabs/disks damped with a solvent were used to extract target compounds such as pesticides from fruit surfaces by swabbing the surface of multiple fruit pieces of a sample batch before ambient MS analysis (Edison et al. 2011; Edison, Lin, and Parrales 2011). Moreover, signals between repeated measurements of a given sample often show high fluctuation during the ambient MS surface analysis, but the correct use of ratio of ions, as well as the normalization of the responses by using internal standards can correct the poor repeatability observed. Although this relatively poor reproducibility can be acceptable for screening analysis, it may be critical for quantitative analysis as well as for the comparison between groups of a high number of samples for fingerprinting approaches such as food authentication or food traceability.

In order to improve LODs, high-resolution mass analyzers (i.e., TOF and Orbitrap) have been used, taking advantage of both the high sensitivity and selectivity of these instruments. This approach allows LODs to be achieved below the maximum residue levels (MRL) established for a large group of compounds in food safety and fulfilling the legislation in environmental analysis. However, efforts are still devoted to detecting analytes at trace levels when dealing with highly complex samples (biological, environmental, food or even small organisms), sample manipulation strategies being necessary in order to obtain satisfactory LODs. The choice of the sample manipulation largely depends on the amount of matrix interferences, the sensitivity of the compounds of interest and the selectivity required for the analytical objective.

In the last five years, the use of ambient MS techniques in metabolomic studies has exponentially increased due to their potential for providing a wealth chemical information, for their high capacity to identify potential markers and for their capability to classify samples based on the ambient MS profile as fingerprint (Clendinen, Monge, and Fernández 2017). Targeted and untargeted approaches have been explored for plant biology and agricultural studies, DESI, leaf spray, DART, LAESI and PESI being some of the ambient MS techniques used. Good results were obtained for most of these techniques for classification and discrimination analysis using multivariate statistical approaches such as principal component analysis (PCA). It must be mentioned that laser-based techniques such as LAESI were found to be challenging for plant metabolomics since the ablation process of the material was highly affected by the water content of the sample, so that the data obtained must be carefully evaluated in order to get reliable conclusions (Etalo et al. 2015). Regarding foodomics, much attention has been given to food quality and safety as well as to genetic-modified food. By far, DART-MS has been the most used ambient MS technique in this field, and it has been implemented within targeted metabolomics workflows, including metabolomics fingerprinting for beer origin recognition, authentication of milk products, differentiation between organic and conventional fruits and for olive oil authenticity assessment. However, untargeted approaches are still a challenge, and the extraction efficiency of metabolites of interest needs further improvement for better sensitivity. Furthermore, sample preparation protocols that involve metabolite extraction or sample grinding into powder before analysis have a drastic impact on the compounds that can be detected.

Regarding quantitative analysis, accurate methods are necessary to ensure that the concentration of the contaminants detected in environmental and food safety analysis does not exceed the maximum legislated levels to survey both the risk on human health and the environmental quality standards. There is still some debate about whether ambient MS techniques are suitable for quantification and if they are able to guarantee the absence of false positive and negative results. Among the quantification methods developed with ambient MS techniques, those applied to liquid-based samples showed the most accurate quantification results (Räsänen et al. 2014; Black, Chevallier, and Elliott 2016; Chen et al. 2017). This fact is related to the intrinsic difficulty in obtaining reliable results *via* direct surface analysis, and most of the limitations are related to the representativeness of the sample with respect to the entire sample batch. Moreover, the use of internal standards does not allow this problem to be solved since it is not possible to guarantee uniform spiking on the sample surface. In this context, sample manipulation is needed in order to get accurate and reproducible quantitative results and opens the possibility for matrix-matched calibration when dealing with complex samples.

10.6 DIFFERENT TECHNIQUES

10.6.1 Desorption Electrospray Ionization

Desorption electrospray ionization (DESI) has received significant attention as a mass spectrometry ionization technique, with a variety of experimental configurations for the analysis of foodstuffs and environmental samples (Asimakopoulos and Gonzalez 2019). The DESI method was developed relatively recently in 2004 by the group of G. Cooks (Cooks et al. 2006). Consequently, DESI has not yet revealed its full potential, but it is developing in all directions of scientific research. The technique is based on electrospray ionization under ambient conditions and resembles conventional electrospray (ESI) (Cooks et al. 2006).

In DESI, a mixture of water and organic solvent (e.g., methanol and acetonitrile) is sprayed towards the sample or the surface that contains the sample. As a result, secondary gas-phase ions are produced from the surface of the sample as the charged aqueous droplets, gas-phase ions and metastable atoms impact it (Asimakopoulos and Gonzalez, 2019). The production of these secondary ions is produced by known ESI mechanisms of direct ion emission or total evaporation of the neutral solvent molecules. In addition, momentum transfer from impacting droplets contributes to the desorption of larger molecules (e.g., proteins), along with electrostatic repulsion forces that are created upon impact due to the charge accumulation on the sample surface. The spray conditions of DESI can be tuned for either dry ion or droplet impact by adjusting the distance to the sample or the nebulizing gas flow rate (Asimakopoulos and Gonzalez, 2019). Two proposed mechanisms of DESI ionization are generally accepted. For high–molecular weight compounds (e.g., peptides, carbohydrates and proteins), the DESI spectra contain multiple charged ions indicating a potential mechanism of ESI. For low–molecular weight compounds the ionization mechanism is similar to the atmospheric pressure chemical ionization (APCI), where single charged ions are more prone to be abundant in the mass spectra. The ionization mechanisms in DESI are linked to sample surface properties, such as conductivity and polarity (Asimakopoulos and Gonzalez, 2019). In DESI, quantitation is usually tempered by matrix effects, which can vary significantly between chemicals and samples analyzed (Asimakopoulos and Gonzalez, 2019). Internal standards are often used to increase accuracy and precision but demonstrate setbacks when used for solid samples. DESI can simply ionize large molecules and protonate other analytes to generate ions that derive from compounds that have low desorption energies (high tendency to desorb) such as volatile and semivolatile compounds (e.g., pesticides), low–molecular weight non-polar chemicals (e.g., lipids), low–molecular weight polar compounds (e.g., amino acids) and larger inorganic ions (e.g., perchlorate) (Asimakopoulos and Gonzalez, 2019). DESI is classified as a soft-ionization technique and has been applied to foodstuffs and environmental samples with the aim to screen at high-throughput rates with high sensitivity and specificity.

The range of DESI applications is wide, including assessment of solid surfaces, high-throughput examination of dried solutions (after transferring them onto a solid surface), surface imaging for foodstuff recognition, quality assessment of foodstuff products and other environmental studies. Thus far, DESI has been tailored to a variety of MS detectors (Asimakopoulos and Gonzalez, 2019), including triple quadrupoles, linear ion traps, orbitrap, quadrupole time-of-flight (QTOF), ion mobility/TOF and ion mobility/QTOF hybrids, Q-traps, Fourier transform ion cyclotron resonance (FT-ICR) instruments and even miniature ion trap mass spectrometers.

10.6.2 Desorption Atmospheric Pressure Photoionization

Desorption atmospheric pressure photoionization (DAPPI) is an ambient mass spectrometry technique that uses photoionization in the ionization process, and is therefore highly efficient in the ionization of neutral and nonpolar compounds (Haapala et al. 2007).

DAPPI uses a heated microchip, which delivers hot, vaporized solvent and nebulizer gas to the sampling surface. The hot vapor flow causes the thermal desorption of the analytes from the sample surface, which are then ionized in the gas-phase by photon-initiated reactions. A vacuum ultraviolet (VUV) krypton discharge lamp is used to deliver 10 eV photons, which can ionize any compounds with ionization energies (IEs) below the energy of the photons. The ionization reactions are similar to those in atmospheric pressure photoionization (APPI): the initial reaction is the photoionization of the solvent, i.e., the dopant. The dopant is a solvent that has IE below 10 eV, and which can therefore readily be ionized by the VUV lamp photons. In photoionization, the dopant forms molecular ions (or protonated molecules in the case of acetone), which can react further with the analytes through charge exchange or proton transfer. Typically, analytes that have high PAs (proton affinity) tend to form protonated molecules, while compounds with low PAs are ionized *via* the charge exchange route. In negative ion DAPPI, the analytes can form negative molecular ions, deprotonated molecules or oxidation products. Typical DAPPI (and APPI) dopants are e.g., toluene, acetone or anisole.

Similarly to APPI, DAPPI can be used to ionize completely nonpolar compounds, and the polarity range that can be achieved by DAPPI is therefore wider than with e.g., DESI. The thermal desorption is highly efficient for small molecules, but it prevents the analysis of large, low-volatility and/or thermolabile compounds. Thus so far, the above-described DAPPI setup has been applied to the analysis of pharmaceuticals in tablets (Haapala et al. 2007), drugs of abuse in powder, tablet, resin, plant and paper form, drug metabolites and endogenic steroids from urine (Kauppila, 2019), pesticides in plant material, PAHs in soil, lipids from capsules, butter, milk and food oils, explosives, insect defense compounds from insect skin, and mouse tissue (Kauppila, 2019).

10.6.3 Easy Ambient Sonic-Spray Ionization

The easy ambient sonic-spray ionization (EASI)-MS technique, developed by Eberlin and coworkers in 2006, is certainly the simplest ambient mass spectrometry ionization technique since no temperature or voltage is required to promote the ionization of the molecules. Additionally, an EASI source can be easily constructed in a mass spectrometry lab without any special requirement, and therefore it is low-cost. Due to its high versatility, the EASI source can be adapted to any mass spectrometer, from a single quadrupole to a high-resolution ICR or Orbitrap.

Structurally, an EASI source is constructed from a "T" capillary connection, in which the solvent is pumped through one end and the gas flows through the second; both solvent and air go out through the third end in a high velocity, enabling the desorption and sample ionization. Since the EASI source can be easily hand-constructed, the optimization of the ionization source is very important to reach the best performance of the sonic-spray (Alberici et al. 2019). In general, a polar solvent, such as methanol, is pumped at 20 μL/min and the gas, commonly N_2, but even air, flows at 3 L/min.

Regarding the ionization mechanism, it is based on the sonic-spray ionization (SSI) process, described as the first voltage-, radiation- and temperature-free ionization technique by Hirabayashi and coworkers in 1994 (Hirabayashi et al. 1994).

The high-velocity flow of gas and solvent generates high shearing force, which promotes the formation of very tiny droplets with an unbalanced distribution of cations and anions. Thus, these minute droplets consist of a low-density bipolar stream, enough for molecule ionization. The ionization itself occurs when the solvent evaporates, the droplet collapses and the charges are transferred to the molecules, making available the ionized molecules into the gas-phase. A desorption step occurs before the ionization process, in which the bipolar stream droplets strike the analyzed sample, on a surface, desorbing the molecules from the sample support by a droplet pick-up process.

After EASI ionization, analytes can be analyzed by both positive and negative mass spectrometry, EASI(+) and EASI(–), respectively. The most common EASI(+) ionization is cationization, by means of the formation of cation adducts, mainly with sodium, [M+Na]+, potassium, [M+K]+, and ammonium, [M+NH4]+, although the common protonation, [M+H]+, occurs as well. The deprotonation ionization is the most common process in the EASI(–), however, anionization can also happen. The molecular ion formed depends directly on the molecule structure and deserves to be studied in each case.

The most important feature of the EASI ionization is its ability to analyze intact molecules, with no fragmentation due to soft molecule ionization as a result of the low charge density. The bipolar stream enables the analysis of both EASI(+) and EASI(–) in the same experiment, excluding the necessity of acidified or basified solvents. All these features are advantages that make possible the analysis of complex mixtures with no, or reduced sample preparations.

10.6.4 Direct Analysis in Real Time

Direct analysis in real time (DART) is a mass spectrometric atmospheric pressure ion source. In that ion source, solids, liquids and gases ionize in open air under ambient conditions. DART has been used as a commercial product since 2005 (Cody et al. 2005a; Laramee and Cody 2006). DART can work the atmospheric pressure interactions of long lived electronic excited-state atoms or vibronic excited-state molecules with the sample and atmospheric gases (Cody et al. 2005b).

The working principle of the DART ion source is that a gas flows through an enclosed chamber and generates ions, electrons, and excited-state neutral species (atoms and molecules). The gases such as helium, argon and nitrogen can be used. The enclosed chamber has intermediate lenses or grids. Moreover, the excited-state species pass through but neutral gas atoms/molecules remain in that chamber (Cody 2009). The grid electrode which has positioned to that chamber helps to remove ions with opposite polarity to prevent signal loss by ion–ion interaction and ion–electron recombination (Cody et al. 2005b).

The DART ion source can be formed in both positive and negative mode. These ions depend on the nature of the gas, ion polarity and whether dopants are present (Cody et al. 2005a). In positive ion mode, when nitrogen is used, molecular ions (M+.) are formed for low-polarity or nonpolar molecule compounds while protonated [M + H]+ cations are typically observed when helium is used. When an ammonia source is present nearby the DART source adducts have also been observed [M + NH4]+. In negative ion mode, mass spectra are mainly dominated by deprotonated [M – H]– anions for most compounds while some negative-charge ions (M–.) are observed for specific compounds. Other adducts, such as [M + Cl]–, are observed when a suitable dopant is used (Cody et al. 2005a).

Several ionization mechanisms in DART have been presented by Cody et al. (2005a).

10.6.5 Low-Temperature Plasma Ionization

The term "plasma" refers to gases with a significant proportion of particles in a non-neutral state (Langmuir 1928). The plasma is "non-thermal" if the plasma temperature is about ambient (below ~40°C) (Rutscher 2008). In non-thermal plasmas (NTPs), the electron temperature is much higher than the ion and neutral gas temperature. As a consequence, the temperature of the plasma system is near the ambient temperature.

In mass spectrometry, non-thermal plasmas are commonly called "low-temperature plasma" (LTP), and this term will be used in this paragraph. It wasn't until 2008 that its potential was recognized as an ion source for instrumental analysis, especially for mass spectrometry (MS) (Harper et al. 2008).

Many analytical applications have been reported since 2008, because LTP works for solid, liquid and gaseous samples, is suitable for imaging, has low technical requirements and is capable of ionizing a wide range of chemically diverse compounds (Harper et al. 2008).

Four important aspects can be identified: (1) plasma generation and plasma processes, (2) ionization and fragmentation of molecules, (3) desorption and surface reactions and (4) ion transfer.

10.6.6 Plasma-Assisted Desorption Ionization

Plasma-assisted desorption ionization (PADI) is an atmospheric pressure radio-frequency (RF) plasma-based technique that can be coupled with atmospheric pressure sampling mass spectrometry. One of the main advantages of plasma-based desorption/ionization techniques are that they are able to yield mass spectral information from a range of analyte types without the need for surface/sample preparation. PADI was one of the original plasma-based sources developed for ambient mass spectrometry and was first described for use as a desorption/ionization source for ambient mass spectrometry by Ratcliffe et al. (Ratcliffe et al. 2007); however the PADI source was based upon the RF plasma needle developed by Stoffels et al. (Stoffels et al. 2002), and was originally designed for biomedical applications. PADI, however, has lost favor in recent years due to the number of other low-temperature plasma sources developed, but still has great potential, especially in the areas of analyte detection for food and environmental applications.

10.6.7 Extractive Electrospray Ionization

Extractive electrospray ionization (EESI) mass spectroscopy is a dual-spray powerful and versatile analytical technique which attracts much interest due to its advantages over conventional electrospray ionization (ESI). It is a powerful ambient ionization technique that can provide comprehensive mass spectrometric (MS) information for many samples with complex mixtures, aerosols, complex liquids or suspensions without any sample pretreatment and fast testing. It is widely used in many fields requiring high-throughput analysis, such as online detection of chemical reaction products and intermediates, detection of native biomolecules environmental monitoring in the field and in-vivo metabolomics. It has also been employed for the rapid characterization of living objects' native proteins, and metabolic biomarkers proteomics (analysis of proteins) and petroleomics (analysis of petroleum) (Mital 2019).

EESI mass spectroscopy, a technique derived from ESI, was first introduced by Cooks et al. in 2006 (Cooks et al. 2006) as a powerful and versatile analytical technique. In the traditional ESI process, the analytes are dissolved in solvent and delivered through a capillary. A very high voltage is applied at the tip of the capillary, which charges the molecules in the solvent. These highly charged molecules, when exiting the capillary tip and pushing towards the evaporation chamber, form the cone shape (Taylor cone) initially, which further is converted to a fine spray as droplets burst away due to the repulsion among the like charges. Thus, a plume of charged droplets is formed with the help of a high electric field and sometimes assisted with a sheath glass. The droplets in the spray go through a series of divisions because the solvent within these droplets gradually evaporates (with the assistance of nitrogen gas pumped into the chamber), forcing the charges in the molecules within these droplets (which, again, are identical) closer together. When these ions are pushed close enough together, they will repel each other (this behavior is known as the Coulomb force), causing the droplets to divide into two smaller droplets.

10.6.8 Electrospray Laser Desorption Ionization

Electrospray laser desorption ionization (ELDI) combines the features of both ESI (electrospray ionization) and LD (laser desorption). It allows direct, sensitive and rapid analyses of chemical compounds on sample surfaces (Huang et al. 2011).

Both small organic molecules and large biomolecules, such as proteins, are readily desorbed by a pulse laser operated at a UV wavelength (266 nm) for a duration of nanoseconds. Peptides and proteins in dry or wet media, such as blood, serum, saliva or tears, and the chemical components on the surfaces of various solids, such as animal brain tissue, have been characterized directly through ELDI-MS without sample pretreatment.

Chemical and biological compounds dissolved in polar and nonpolar solvents can also be characterized directly by ELDI-MS through the assistance of fine carbon powders, which absorb the laser energy, suspended in the solution.

10.6.9 Sorptive Tape Extraction Laser Desorption Ionization

In sorptive tape extraction (STE) a sorbent surface is used for molecular imprisonment and posterior instrument analysis. It is a sorption-based sampling technique in which a flexible polydimethylsiloxane (PDMS) tape is used to recover analytes at the surface of a solid matrix by direct contact as well as from the headspace in equilibrium with it (Bicchi et al. 2007). Solutes enriched on the inert PDMS material can be recovered either by solvent desorption or by thermo-desorption.

10.6.10 Rapid Evaporative Ionization Mass Spectrometry

The production of ions of organic compounds upon heat treatment of solid samples has been known since the 1970s (Lebedev 2015).

This phenomenon was used in the development of a direct (desorption) chemical ionization technique based on rapid heating of the sample to a temperature at which evaporation proceeds faster than thermal degradation. In open-air conditions, the thermal energy of a species efficiently decreases owing to collisions with gas molecules. This kind of cooling is characterized by additional suppression of thermal degradation.

For rapid evaporative ionization mass spectrometry (REIMS) one can use Joule heating, contact heating or radiative heating. For instance, the mass spectra of biological tissues obtained using laser ablation (laser desorption ionization) and by the electrochemical Joule heating method are very similar.

Although there are various applications of REIMS, the most in-demand application is still the analysis of biological tissues, including in-vivo analysis which is of crucial importance for medicine. Experiments showed that REIMS of biological tissues provides information of the molecular masses of many biologically important groups of compounds (first of all, lipids) (Schäfer et al. 2009).

Possible heaters may be commercially available surgical instruments, viz., surgical lasers and electrosurgical instruments.

10.7 CONCLUDING REMARKS

Considering the continuous developments regarding the technology field, predicting the directions of ambient MS techniques in food and environmental applications is still a challenging task. However, the attractive features of these techniques such as real time and *in-situ* analysis, low sample requirements, high throughput and minimal sample preparation will clearly boost the expansion and the use of ambient MS in environmental and food analytical laboratories.

Today, most of the new ambient MS applications published in food and environmental analysis are still homemade prototypes for proof-of-concept studies. Thus, more work needs to be done to understand which ambient MS techniques have the capabilities for the development of real routine analytical methods. Until now, DESI, DART, EESI, PS and LAESI have shown the best capabilities to meet the demands of modern analytical laboratories. Furthermore, the continuous effort on the automatization and robotization of the initial prototypes will improve the reproducibility necessary for some important food applications such as food authenticity and food profiling. On the other hand, it is clearly demonstrated that ambient MS techniques are very attractive for the screening of a large number of compounds. Thus, the development of multimode and/or hybrid ambient MS techniques would allow concurring different ionization mechanisms in order to expand the simultaneous detection of a broader spectrum of chemical species, obtaining additional chemical information of a sample in a single experiment. Although no additional sample manipulation other than the *in-situ* sample processing is necessary in most applications, the matrix complexity is still a challenge. Hence, the combination of ambient MS techniques with well-established simple and fast sample pretreatment procedures may be necessary for reducing both the matrix interferences and matrix effects. The combination of ambient MS techniques with in line and/or *in-situ* sample processing procedures should be considered to overcome these problems. Moreover, the need to detect contaminants at trace levels and the accurate surface quantification remain some of the most important difficulties in ambient MS analysis. Finally, the development of hand-portable mass spectrometers using ambient ionization sources would greatly enhance the *in-situ* analysis of food and environmental samples directly in the field.

ACKNOWLEDGMENTS

The authors gratefully acknowledge the financial support received from Spanish Ministry of Science, Innovation and Universities under the project PGC2018-095013-B-I00 and from the Agency for Administration of University and Research Grants (Generalitat de Catalunya, Spain) under the project 2017SGR-310.

REFERENCES

Alberici, R. M., G. D. Fernandes, and D. Barrera-Arellano. 2019. *Easy Ambient Sonic-Spray Ionization Mass Spectrometry in Food Analysis and Foodomics in Ambient Mass Spectrometry Techniques in Food and Environment*, edited by L. M. L. Nollet & B. K. Munjanja. Boca Raton, FL: CRC Press.

Andrade, F. J., J. T. Shelley, W. C. Wetze, M. R. Webb, G. Gamez, S. J. Ray, and G. M. Hieftje. 2008. "Atmospheric Pressure Chemical Ionization Source. 1. Ionization of Compounds in the Gas Phase." *Analytical Chemistry* 80 (8): 2646–53.

Asimakopoulos, A. G. and S. V. Gonzalez. 2019. *Desoprtion Electrospray Ionization in Ambient Mass Spectrometry Techniques in Food and Environment*, edited by L. M. L. Nollet & B. K. Munjanja. Boca Raton, FL: CRC Press.

Avula, B., T. J. Smillie, Y. H. Wang, J. Zweigenbaum, and I. A. Khan. 2015. "Authentication of True Cinnamon (Cinnamon Verum) Utilising Direct Analysis in Real Time (DART)-QToF-MS." *Food Additives and Contaminants: Part A* 32 (1): 1–8.

Badu-Tawiah, A. K., L. S. Eberlin, Z. Ouyang, and R. G. Cooks. 2013. "Chemical Aspects of the Extractive Methods of Ambient Ionization Mass Spectrometry." *Annual Review of Physical Chemistry* 64 (1): 481–505.

Bai, Y., J. Zhang, Y. Bai, and H. Liu. 2012. "Direct Analysis in Real Time Mass Spectrometry Combined with Single-Drop Liquid-Liquid-Liquid Microextraction for the Rapid Analysis of Multiple Phytohormones in Fruit Juice." *Analytical and Bioanalytical Chemistry* 403 (8): 2307–14.

Balog, J., D. Perenyi, C. Guallar-Hoyas, A. Egri, S. D. Pringle, S. Stead, O. P. Chevallier, C. T. Elliott, and Z. Takats. 2016. "Identification of the Species of Origin for Meat Products by Rapid Evaporative Ionization Mass Spectrometry." *Journal of Agricultural and Food Chemistry* 64 (23): 4793–800.

Beach, D. G., C. M. Walsh, P. Cantrell, W. Rourke, S. O'Brien, K. Reeves, and P. McCarron. 2016. "Laser Ablation Electrospray Ionization High-Resolution Mass Spectrometry for Regulatory Screening of Domoic Acid in Shellfish." *Rapid Communications in Mass Spectrometry* 30 (22): 2379–87.

Benassi, M., A. Berisha, W. Romão, E. Babayev, A. Römpp, and B. Spengler. 2013. "Petroleum Crude Oil Analysis Using Low-Temperature Plasma Mass Spectrometry." *Rapid Communications in Mass Spectrometry* 27: 825–34.

Berchtold, C., V. Müller, L. Meier, S. Schmid, and R. Zenobi. 2013. "Direct Detection of Chlorpropham on Potato Skin Using Desorption Electrospray Ionization." *Journal of Mass Spectrometry* 48 (5): 587–93.

Berkel, G. J. V., A. D. Sanchez, and J. M. E. Quirke. 2002. "Thin-Layer Chromatography and Electrospray Mass Spectrometry Coupled Using a Surface Sampling Probe." *Analytical Chemistry* 74 (24): 6216–23.

Bicchi C., C. Cordero, E. LIberto, P. Rubiolo, B. Sgorbini, and P. Sandra. 2007. "Sorptive Tape Extraction in the Analysis of the Volatile Fraction Emitted from Biological Solid Matrices." *Journal of Chromatography A* 1148 (2): 137–44.

Black, C., O. P. Chevallier, and C. T. Elliott. 2016. "The Current and Potential Applications of Ambient Mass Spectrometry in Detecting Food Fraud." *Trends in Analytical Chemistry* 82: 268–78.

Bodzon-Kulakowska, A., A. Drabik, J. Ner, J. H. Kotlinska, and P. Suder. 2014. "Desorption Electrospray Ionisation (DESI) for Beginners - How to Adjust Settings for Tissue Imaging." *Rapid Communications in Mass Spectrometry* 28: 1–9.

Boone, E. J., A. Laskin, J. Laskin, C. Wirth, P. B. Shepson, B. H. Stirm, and K. A. Pratt. 2015. "Aqueous Processing of Atmospheric Organic Particles in Cloud Water Collected via Aircraft Sampling." *Environmental Science and Technology* 49: 8523–30.

Cain, J., A. Laskin, M. R. Kholghy, M. J. Thomson, and H. Wang. 2014. "Molecular Characterization of Organic Content of Soot Along the Centerline of a Coflow Diffusion Flame." *Physical Chemistry Chemical Physics* 16 (47): 25862–75.

Cajka, T., K. Riddellova, M. Tomaniova, and J. Hajslova. 2011. "Ambient Mass Spectrometry Employing a DART Ion Source for Metabolomic Fingerprinting/ Profiling: A Powerful Tool for Beer Origin Recognition." *Metabolomics* 7: 500–508.

Cajka, T., K. Riddellova, P. Zomer, H. Mol, and J. Hajslova. 2011. "Direct Analysis of Dithiocarbamate Fungicides in Fruit by Ambient Mass Spectrometry." *Food Additives and Contaminants: Part A* 28 (10): 1372–82.

Chan, C. C., M. S. Bolgar, S. A. Miller, and A. B. Attygalle. 2010. "Desorption Ionization by Charge Exchange (DICE) for Sample Analysis Under Ambient Conditions by Mass Spectrometry." *Journal of the American Society for Mass Spectrometry* 21 (9): 1554–60.

Chen, H., A. Venter, and R. G. Cooks. 2006. "Extractive Electrospray Ionization for Direct Analysis of Undiluted Urine, Milk and Other Complex Mixtures Without Sample Preparation." *Chemical Communications* 42 (19): 2042–44.

Chen, H., A. Wortmann, and R. Zenobi. 2007. "Neutral Desorption Sampling Coupled to Extractive Electrospray Ionization Mass Spectrometry for Rapid Differentiation of Biosamples by Metabolomic Fingerprinting." *Journal of Mass Spectrometry* 42: 1123–35.

Chen, R., J. Deng, L. Fang, Y. Yao, B. Chen, X. Wang, and T. Luan. 2017. "Recent Applications of Ambient Ionization Mass Spectrometry in Environmental Analysis." *Trends in Environmental Analytical Chemistry* 15: 1–11.

Clendinen, C. S., M. E. Monge, and F. M. Fernández. 2017. "Ambient Mass Spectrometry in Metabolomics." *Analyst* 142: 3101–17.

Cody, R. B. 2009. "Observation of Molecular Ions and Analysis of Nonpolar Compounds with Direct Analysis in Real Time Ion Source." *Analytical Chemistry* 81: 1101–1107.

Cody, R. B., and A. J. Dane. 2015. "Direct Analysis in Real Time (DART)." In *Ambient Ionization Mass Spectrometry*, edited by M. Domin and R. Cody, 23–57. London, UK: Royal Society of Chemistry.

Cody, R. B., Laramee, J. A., Durst, H. D. 2005a. "Versatile New Ion Source for the Analysis of Materials in Open Air Under Ambient Conditions." *Analytical Chemistry* 77: 2297–302.

Cody, R. B., J. A. Laramee, and H. D. Durst. 2005b. "Direct Analysis in Real Time Mass spectrometry." *JEOL News* 40: 8–12.

Cody, R. C. 2013. "What Is the Opposite of Pandora's Box? Direct Analysis, Ambient Ionization, and a New Generation of Atmospheric Pressure Ion Sources." *Mass Spectrometry* 2: 7–10.

Cooks, R. G., and T. Mueller. 2013. "Through a Glass Darkly: Glimpses into the Future of Mass Spectrometry." *Mass Spectrometry* 2: 1–10.

Cotte-Rodríguez, I., and R. G. Cooks. 2006. "Non-Proximate Detection of Explosives and Chemical Warfare Agent Simulants by Desorption Electrospray Ionization Mass Spectrometry." *Chemical Communications* 28 (28): 2968–70.

Crawford, E., and B. Musselman. 2012. "Evaluating a Direct Swabbing Method for Screening Pesticides on Fruit and Vegetable Surfaces Using Direct Analysis in Real Time (DART) Coupled to an Exactive Benchtop Orbitrap Mass Spectrometer." *Analytical and Bioanalytical Chemistry* 403: 2807–12.

Danhelova, H., J. Hradecky, S. Prinosilova, T. Cajka, K. Riddellova, L. Vaclavik, and J. Hajslova. 2012. "Rapid Analysis of Caffeine in Various Coffee Samples Employing Direct Analysis in Real-Time Ionization-High-Resolution Mass Spectrometry." *Analytical and Bioanalytical Chemistry* 403: 2883–89.

Deng, J., T. Yu, Y. Yao, Q. Peng, L. Luo, B. Chen, X. Wang, Y. Yang, and T. Luan. 2017. "Surface-Coated Wooden-Tip Electrospray Ionization Mass Spectrometry for Determination of Trace Fluoroquinolone and Macrolide Antibiotics in Water." *Analytica Chimica Acta* 954: 52–59.

Ding, X., and Y. Duan. 2015. "Plasma-Based Ambient Mass Spectrometry Technniques: The Current Status and Future Prospective." *Mass Spectrometry Reviews* 34: 449–73.

Domin, M., and R. Cody (Ed.). 2015. *Ambient Ionization Mass Spectrometry*. Cambridge, UK: RSC Publishing.

Edison, S. E., L. A. Lin, B. M. Gamble, J. Wong, and K. Zhang. 2011. "Surface Swabbing Technique for the Rapid Screening for Pesticides Using Ambient Pressure Desorption Ionization with High-Resolution Mass Spectrometry." *Rapid Communications in Mass Spectrometry* 25 (1): 127–39.

Edison, S. E., L. A. Lin, and L. Parrales. 2011. "Practical Considerations for the Rapid Screening for Pesticides Using Ambient Pressure Desorption Ionisation with High-Resolution Mass Spectrometry." *Food Additives and Contaminants: Part A* 28 (10): 1393–404.

Espy, R. D., M. Wleklinski, X. Yan, and R. G. Cooks. 2014. "Beyond the Flask: Reactions on the Fly in Ambient Mass Spectrometry." *Trends in Analytical Chemistry* 57: 135–46.

Etalo, D. W., C. H. de Vos, M. H. A. J. Joosten, and R. D. Hall. 2015. "Spatially-Resolved Plant Metabolomics: Some Potentials and Limitations of Laser-Ablation Electrospray Ionization (LAESI) Mass Spectrometry Metabolite Imaging." *Plant Physiology* 169: 1424–35.

Evard, H., A. Kruve, R. Lõhmus, and I. Leito. 2015. "Paper Spray Ionization Mass Spectrometry: Study of a Method for Fast-Screening Analysis of Pesticides in Fruits and Vegetables." *Journal of Food Composition and Analysis* 41: 221–25.

Ewing, K. J., D. Gibson, J. Sanghera, and F. Miklos. 2015. "Desorption Electrospray Ionization-Mass Spectrometric Analysis of Low Vapor Pressure Chemical Particulates Collected from a Surface." *Analytica Chimica Acta* 853: 368–74.

Fang, L., J. Deng, Y. Yu, Y. Yang, X. Wang, H. Liu, and T. Luan. 2016. "Coupling Liquid-Phase Microextraction with Paper Spray for Rapid Analysis of Malachite Green, Crystal Violet and Their Metabolites in Complex Samples Using Mass Spectrometry." *Analytical Methods* 8 (36): 6651–56.

Farré, M., Y. Picó, and D. Barceló. 2013. "Direct Peel Monitoring of Xenobiotics in Fruit by Direct Analysis in Real Time Coupled to a Linear Quadrupole Ion Trap-Orbitrap Mass Spectrometer." *Analytical Chemistry* 85 (5): 2638–44.

Garcia-Reyes, J. F., A. U. Jackson, A. Molina-Diaz, and R. G. Cooks. 2009. "Desorption Electrospray Ionization Mass Spectrometry for Trace Analysis of Agrochemicals in Food." *Analytical Chemistry* 81 (2): 820–29.

Garcia-Reyes, J. F., F. Mazzoti, J. D. Harper, N. A. Charipar, S. Oradu, Z. Ouyang, G. Sindona, and R. G. Cooks. 2009. "Direct Olive Oil Analysis by Low-Temperature Plasma (LTP) Ambient Ionization Mass Spectrometry." *Rapid Communications in Mass Spectrometry* 23: 3057–62.

Garrett, R., C. M. Rezende, and D. R. Ifa. 2013. "Coffee Origin Discrimination by Paper Spray Mass Spectrometry and Direct Coffee Spray Analysis." *Analytical Methods* 5 (21): 5944.

Gerbig, S., and Z. Takáts. 2010. "Analysis of Triglycerides in Food Items by Desorption Electrospray Ionization Mass Spectrometry." *Rapid Communications in Mass Spectrometry* 24: 2186–92.

Gerbig, S., H. E. Brunn, B. Spengler, and S. Schulz. 2015. "Spatially Resolved Investigation of Systemic and Contact Pesticides in Plant Material by Desorption Electrospray Ionization Mass Spectrometry Imaging (DESI-MSI)." *Analytical and Bioanalytical Chemistry* 407 (24): 7379–89.

Gerbig, S., S. Neese, A. Penner, B. Spengler, and S. Schulz. 2017. "Real-Time Food Authentication Using a Miniature Mass Spectrometer." *Analytical Chemistry* 89 (20): 10717–25.

Girod, M., E. Moyano, D. I. Campbell, and R. G. Cooks. 2011. "Accelerated Bimolecular Reactions in Microdroplets Studied by Desorption Electrospray Ionization Mass Spectrometry." *Chemical Science* 2: 501–10.

Green, F. M., T. L. Salter, I. S. Gilmore, P. Stokes, and G. O'Connor. 2010. "The Effect of Electrospray Solvent Composition on Desorption Electrospray Ionisation (DESI) Efficiency and Spatial Resolution." *Analyst* 135 (4): 731.

Gross, J. H. 2017. "Ambient Desorption/Ionization." In *Mass Spectrometry. A Textbook*, edited by Jürgen H. Gross, 3rd ed., 779–829. Springer.

Guo, Y., Z. Gu, X. Liu, J. Liu, M. Ma, B. Chen, and L. Wang. 2017. "Rapid Analysis of Corni Fructus Using Paper Spray-Mass Spectrometry." *Phytochemical Analysis* 28 (4): 344–50.

Haapala, M., J. Pól, V. Saarela, V. Arvola, T. Kotiaho, R. A. Ketola, S. Franssila, T. J. Kauppila, and R. Kostiainen. 2007. "Desorption Atmospheric Pressure Photoionization." *Analytical Chemistry* 79 (20): 7867–72.

Haddad, R., R. Sparrapan, M. N. Eberlin. 2006. "Desorption Sonic Spray Ionization for (High) Voltage-Free Ambient Mass Spectrometry." *Rapid Communications in Mass Spectrometry* 20: 2901–905.

Hajslova, J., T. Cajka, and L. Vaclavik. 2011. "Challenging Applications Offered by Direct Analysis in Real Time (DART) in Food-Quality and Safety Analysis." *Trends in Analytical Chemistry* 30 (2): 204–18.

Hartmanova, L., V. Ranc, B. Papouskova, P. Bednar, V. Havlicek, and K. Lemr. 2010. "Fast Profiling of Anthocyanins in Wine by Desorption Nano-Electrospray Ionization Mass Spectrometry." *Journal of Chromatography A* 1217 (25): 4223–28.

Hirabayashi, A., M. Sakairi, and H. Koizumi. 1994. "Sonic Spray Ionization Method for Atmospheric Pressure Ionization Mass Spectrometry." *Analytical Chemistry* 66: 4557–59.

Hiraoka, K., K. Nishidate, K. Mori, D. Asakawa, and S. Suzuki. 2008. "Development of Probe Electrospray Using a Solid Needle." *Rapid Communications in Mass Spectrometry* 21: 3139–44.

Hrbek, V., L. Vaclavik, O. Elich, and J. Hajslova. 2014. "Authentication of Milk and Milk-Based Foods by Direct Analysis in Real Time Ionization-High Resolution Mass Spectrometry (DART-HRMS) Technique: A Critical Assessment." *Food Control* 36 (1): 138–45.

Huang, M.-Z., S.-C. Cheng, Y.-T. Cho, and J. Shiea. 2011. "Ambient Ionization Mass Spectrometry: A Tutorial." *Analytica Chimica Acta* 702: 1–15.

Huang, M. Z., S. C. Cheng, S. S. Jhang, C. C. Chou, C. N. Cheng, J. Shiea, I. A. Popov, and E. N. Nikolaev. 2012. "Ambient Molecular Imaging of Dry Fungus Surface by Electrospray Laser Desorption Ionization Mass Spectrometry." *International Journal of Mass Spectrometry* 325–327: 172–82.

Huang, X. Y., X. W. Fang, X. Zhang, X. M. Dai, X. L. Guo, H. W. Chen, and L. P. Luo. 2014. "Direct Detection of Chloramphenicol in Honey by Neutral Desorption-Extractive Electrospray Ionization Mass Spectrometry." *Analytical and Bioanalytical Chemistry* 406 (29): 7705–14.

Janfelt, C. 2015. "Imaging of Plant Materials Using Indirect Desorption Electrospray Ionization Mass Spectrometry." In *Mass Spectrometry Imaging of Small Molecules*, edited by L. He, 91–97. New York, NY: Springer New York.

Javanshad, R., and A. R. Venter. 2017. "Ambient Ionization Mass Spectrometry: Real-Time, Proximal Sample Processing and Ionization." *Analytical Methods* 9 (34): 4896–4907.

Jjunju, F. P. M., S. Maher, D. E. Damon, R. M. Barrett, S. U. Syed, R. M. A. Heeren, S. Taylor, and A. K. Badu-Tawiah. 2016. "Screening and Quantification of Aliphatic Primary Alkyl Corrosion Inhibitor Amines in Water Samples by Paper Spray Mass Spectrometry." *Analytical Chemistry* 88 (2): 1391–400.

Joyce, N. I., C. C. Eady, P. Silcock, N. B. Perry, and J. W. Van Klink. 2013. "Fast Phenotyping of LFS-Silenced (Tearless) Onions by Desorption Electrospray Ionization Mass Spectrometry (DESI-MS)." *Journal of Agricultural and Food Chemistry* 61 (7): 1449–56.

Kalachova, K., J. Pulkrabova, L. Drabova, T. Cajka, V. Kocourek, and J. Hajslova. 2011. "Simplified and Rapid Determination of Polychlorinated Biphenyls, Polybrominated Diphenyl Ethers, and Polycyclic Aromatic Hydrocarbons in Fish and Shrimps Integrated into a Single Method." *Analytica Chimica Acta* 707 (1–2): 84–91.

Kauppila, T. 2019. *Desorption Atmospheric Pressure Photoionization in Ambient Mass Spectrometry Techniques in Food and Environment*, edited by L. M. L. Nollet & B. K. Munjanja. Boca Raton, FL: CRC Press.

Kauppila, T. J., and A. Vaikkinen. 2014. "Ambient Mass Spectrometry: Food and Environmental Applications." In *Fast Liquid Chromatography–Mass Spectrometry*

Methods in Food and Environmental Analysis, edited by O. Núñez, H. Gallart-Ayala, C. P. Martins, and P. Lucci, 271–323. Singapore: Imperial College Press.

Kauppila, T. J., and R. Kostiainen. 2017. "Ambient Mass Spectrometry in the Analysis of Compounds of Low Polarity." *Analytical Methods* 9 (34): 4936–53.

Kertesz, V., and G. J. Van Berkel. 2008. "Improved Imaging Resolution in Desorption Electrospray Ionization Mass Spectrometry." *Rapid Communications in Mass Spectrometry* 22: 2639–44.

Kertesz, V., and G. J. Van Berkel. 2010. "Fully Automated Liquid Extraction-Based Surface Sampling and Ionization Using a Chip-Based Robotic Nanoelectrospray Platform." *Journal of Mass Spectrometry* 45 (3): 252–60.

Klampfl, C. W., and M. Himmelsbach. 2015. "Direct Ionization Methods in Mass Spectrometry: An Overview." *Analytica Chimica Acta* 890: 44–59.

Langmuir, I. 1928. "Oscillations in Ionized Gases." *PNAS* 14 (8): 627–37.

Laramee, J. A., Cody, R. B. 2006. Method for Atmospheric Pressure Analyte Ionization. U.S. Patent 7, 112, 785, September, 26.

Laskin, J., and I. Lanekoff. 2016. "Ambient Mass Spectrometry Imaging Using Direct Liquid Extraction Techniques." *Analytical Chemistry* 88 (1): 52–73.

Laskin, J., B. S. Heath, P. J. Roach, L. Cazares, and O. J. Semmes. 2012. "Tissue Imaging Using Nanospray Desorption Electrospray Ionization Mass Spectrometry." *Analytical Chemistry* 84 (1): 141–48.

Lebedev, A. T. 2015. "Ambient Ionization Mass Spectrometry." *Russian Chemical Reviews* 87 (7): 665–92.

Li, B., N. Bjarnholt, S. H. Hansen, and C. Janfelt. 2011. "Characterization of Barley Leaf Tissue Using Direct and Indirect Desorption Electrospray Ionization Imaging Mass Spectrometry." *Journal of Mass Spectrometry* 46 (12): 1241–46.

Li, X., B. Hu, J. Ding, and H. Chen. 2011. "Rapid Characterization of Complex Viscous Samples at Molecular Levels by Neutral Desorption Extractive Electrospray Ionization Mass Spectrometry." *Nature Protocols* 6 (7): 1010–25.

Liu, C., J. Yang, J. Wang, Y. Hu, W. Zhao, Z. Zhou, F. Qi, and Y. Pan. 2016. "Extractive Atmospheric Pressure Photoionization (EAPPI) Mass Spectrometry: Rapid Analysis of Chemicals in Complex Matrices." *Journal of the American Society for Mass Spectrometry* 27 (10): 1597–605.

Liu, J., H. Wang, R. G. Cooks, and Z. Ouyang. 2011. "Leaf Spray: Direct Chemical Analysis of Plant Material and Living Plants by Mass Spectrometry." *Analytical Chemistry* 83 (20): 7608–13.

Liu, J., N. E. Manicke, X. Zhou, R. G. Cooks, and Z. Ouyang. 2015. "Paper Spray." In *Ambient Ionization Mass Spectrometry*, edited by M. Domin and R. Cody, 389–421. Cambridg, UK: Royal Society of Chemistry.

Lojza, J., T. Cajka, V. Schulzova, K. Riddellova, and J. Hajslova. 2012. "Analysis of Isoflavones in Soybeans Employing Direct Analysis in Real-Time Ionization-High-Resolution Mass Spectrometry." *Journal of Separation Science* 35 (3): 476–81.

Lu, H., H. Zhang, T. Zhu, Y. Xiao, S. Xie, H. Gu, M. Cui, and L. Luo. 2017. "Metabolic Effects of Clenbuterol and Salbutamol on Pork Meat Studied Using Internal Extractive Electrospray Ionization Mass Spectrometry." *Scientific Reports* 7 (1): 1–8.

Luo, L., T. Yu, X. Liu, X. Huang, X. Fang, X. Zhang, and X. Guo. 2017. "Direct Determination of Chlorpyrifos in Honey by Neutral Desorption-Extractive Electrospray Ionization Mass Spectrometry." *Analytical Letters* 50 (12): 1939–49.

Luosujärvi, L., S. Kanerva, V. Saarela, S. Franssila, R. Kostiainen, T. Kotiaho, and T. J. Kauppila. 2010. "Environmental and Food Analysis by Desorption Atmospheric Pressure Photoionization-Mass Spectrometry." *Rapid Communications in Mass Spectrometry* 24: 1343–50.

Mainero-Rocca, L., J. Cecca, N. L'Episcopo, and G. Fabrizi. 2017. "Ambient Mass Spectrometry: Direct Analysis of Dimethoate, Tebuconazole, and Trifloxystrobin on Olive and Vine Leaves by Desorption Electrospray Ionization Interface." *Journal of Mass Spectrometry* 52 (11): 709–19.

Mattarozzi, M., F. Bianchi, M. Milioli, A. Cavazza, and M. Careri. 2016. "An Innovative Method Based on Quick, Easy, Cheap, Effective, Rugged, and Safe Extraction Coupled to Desorption Electrospray Ionization-High Resolution Mass Spectrometry for Screening the Presence of Paralytic Shellfish Toxins in Clams." *Talanta* 147: 416–21.

Mirabelli, M. F., J. C. Wolf, and R. Zenobi. 2016. "Direct Coupling of Solid-Phase Microextraction with Mass Spectrometry: Sub-Pg/g Sensitivity Achieved Using a Dielectric Barrier Discharge Ionization Source." *Analytical Chemistry* 88 (14): 7252–58.

Mital, S.2019. *Extractive Electrospray Ionization in Food Analysis and Foodomics in Ambient Mass Spectrometry Techniques in Food and Environment*, edited by L. M. L. Nollet & B. K. Munjanja. Boca Raton, FL: CRC Press.

Monge, M. E., G. A. Harris, P. Dwivedi, and F. M. Fernández. 2013. "Mass Spectrometry: Recent Advances in Direct Open Air Surface Sampling/Ionization." *Chemical Reviews* 113 (4): 2269–308.

Montowska, M., M. R. Alexander, G. A. Tucker, and D. A. Barrett. 2015. "Authentication of Processed Meat Products by Peptidomic Analysis Using Rapid Ambient Mass Spectrometry." *Food Chemistry* 187: 297–304.

Montowska, M., W. Rao, M. R. Alexander, G. A. Tucker, and D. A. Barrett. 2014. "Tryptic Digestion Coupled with Ambient Desorption Electrospray Ionization and Liquid Extraction Surface Analysis Mass Spectrometry Enabling Identification of Skeletal Muscle Proteins in Mixtures and Distinguishing between Beef, Pork, Horse, Chicken, and T." *Analytical Chemistry* 86 (9): 4479–87.

Moyano, E., and M. T. Galceran. 2015. "Direct Analysis of Pesticides by Stand-Alone Mass Spectrometry: Flow Injection and Ambient Ionization." In *Application in High Resolution Mass Spectrometry*, edited by R. Romero-González and A. Garrido Frenich. Elsevier Inc.

Na, N., C. Zhang, M. Zhao, S. Zhang, C. Yang, X. Fang, and X. Zhang. 2007. "Direct Detection of Explosives on Solid Surfaces by Mass Spectrometry with an Ambient Ion Source Based on Dielectric Barrier Discharge." *Journal of Mass Spectrometry* 42 (8): 1079–85.

Na, N., M. Zhao, S. Zhang, C. Yang, and X. Zhang. 2007. "Development of a Dielectric Barrier Discharge Ion Source for Ambient Mass Spectrometry." *Journal of the American Society for Mass Spectrometry* 18 (10): 1859–62.

Nemes, P., and A. Vertes. 2007. "Laser Ablation Electrospray Ionization for Atmospheric Pressure, in Vivo, and Imaging Mass Spectrometry." *Analytical Chemistry* 79 (21): 8098–106.

Nielen, M. W. F., H. Hooijerink, P. Zomer, and J. G. J. Mol. 2011. "Desorption Electrospray Ionization Mass Spectrometry in the Analysis of Chemical Food Contaminants in Food." *Trends in Analytical Chemistry* 30 (2): 165–80.

Novotná, H., O. Kmiecik, M. Gałazka, V. Krtková, A. Hurajová, V. Schulzová, E. Hallmann, E. Rembiałkowska, and J. Hajšlová. 2012. "Metabolomic Fingerprinting Employing DART-TOFMS for Authentication of Tomatoes and Peppers from Organic and Conventional Farming." *Food Additives and Contaminants: Part A* 29 (9): 1335–46.

Pereira, I., T. C. Caravalho, R Wanderson, P. R. Filgueiras, B. G. Laviola, C. M. Rodrigues, P. V. Abdelnur, and B. G. Vaz. 2017. "Differentiation of Toxic and Non-Toxic Leaves of *Jatropha Curcas* L. Genotypes by Leaf Spray Mass Spectrometry." *Journal of the Brazilian Chemical Society* 28 (8): 1461–66.

Porcari, A. M., G. D. Fernandes, D. Barrera-Arellano, M. N. Eberlin, and R. M. Alberici. 2016. "Food Quality and Authenticity Screening via Easy Ambient Sonic-Spray Ionization Mass Spectrometry." *Analyst* 141 (4): 1172–84.

Porcari, A. M., G. D. Fernandes, K. R. A. Belaz, N. V. Schwab, V. G. Santos, R. M. Alberici, V. A. Gromova, M. N. Eberlin, A. T. Lebedev, and A. Tata. 2014. "High Throughput MS Techniques for Caviar Lipidomics." *Analytical Methods* 6 (8): 2436.

Rajchl, A., L. Drgová, A. Grégrová, H. Čížková, R. Ševčík, and M. Voldřich. 2013. "Rapid Determination of 5-Hydroxymethylfurfural by DART Ionization with Time-of-Flight Mass Spectrometry." *Analytical and Bioanalytical Chemistry* 405 (14): 4737–45.

Räsänen, R., P. Dwivedi, F. M. Fernández, and T. J. Kauppila. 2014. "Desorption Atmospheric Pressure Photoionization and Direct Analysis in Real Time Coupled with Travelling Wave Ion Mobility Mass Spectrometry." *Rapid Communications in Mass Spectrometry* 28 (21): 2325–36.

Ratcliffe, L. V., F. J. M. Rutten, D. A. Barrett, T. Whitmore, D. Seymour, C. Greenwood, Y. Aranda-Gonzalvo, S. Robinson, and M. McCoustra. 2007. "Surface Analysis Under Ambient Conditions Using Plasma-Assisted Desorption/Ionization Mass Spectrometry." *Analytical Chemistry* 79 (16): 6094–6101.

Roach, P. J., J. Laskin, and A. Laskin. 2010a. "Molecular Characterization of Organic Aerosols Using Nanospray-Desorption/Electrospray Ionization-Mass Spectrometry." *Analytical Chemistry* 82 (19): 7979–86.

Roach, P. J., J. Laskin, and A. Laskin. 2010b. "Nanospray Desorption Electrospray Ionization: An Ambient Method for Liquid-Extraction Surface Sampling in Mass Spectrometry." *The Analyst* 135 (9): 2233.

Rutscher, A. 2008. "Characteristics of Low-Temperature Plasmas Under Nonthermal." In M. S. Hippler H; Schmidt (Ed.), *Low Temperature Plasmas: Fundamentals, Technologies and Techniques*. Wiley-VCH.

Sampson, J. S., A. M. Hawkridge, and D. C. Muddiman. 2006. "Generation and Detection of Multiply-Charged Peptides and Proteins by Matrix-Assisted Laser Desorption Electrospray Ionization (MALDESI) Fourier Transform Ion Cyclotron Resonance Mass Spectrometry." *Journal of the American Society for Mass Spectrometry* 17 (12): 1712–16.

Sarkar, P. K., P. K. Prajapati, V. J. Shukla, B. Ravishankar, and A. K. Choudhary. 2009. "Toxicity and Recovery Studies of Two Ayurvedic Preparations of Iron." *Indian Journal of Experimental Biology* 47 (12): 987–92.

Schäfer, K., J. Dénes, K. Albrecht, T. Szaniszló, J. Balog, R. Skoumal, M. Katona, M. Tóth, L. Balogh, and Z. Takáts. 2009. "In Vivo, In Situ Tissue Analysis Using Rapid Evaporative Ionization Mass Spectrometry." *Angewandte Chemie International Edition* 48 (44): 8240–42.

Schurek, J., L. Vaclavik, H. D. Hooijerink, O. Lacina, J. Poustka, M. Sharman, M. Caldow, M. W. F. Nielen, and J. Hajslova. 2008. "Control of Strobilurin Fungicides in Wheat Using Direct Analysis in Real Time Accurate Time-of-Flight and Desorption Electrospray Ionization Linear Ion Trap Mass Spectrometry." *Analytical Chemistry* 80 (24): 9567–75.

Sero, R., O. Nunez, J. Bosch, J. M. Grases, P. Rodriguez, E. Moyano, and M. T. Galceran. 2015. "Desorption Electrospray Ionization-High Resolution Mass Spectrometry for the Screening of Veterinary Drugs in Cross-Contaminated Feedstuffs." *Analytical and Bioanalytical Chemistry*, 1–10.

Sero, R., O. Nunez, and E. Moyano. 2016. "Ambient Ionisation-High Resolution Mass Spectrometry: Environmental, Food, Forensic and Doping Analysis." In *Applications of Time-of-Flight and Orbitrap Mass Spectrometry in Environmental, Food, Doping, and Forensic Analysis*, edited by S. Perez, P. Eichhorn, and D. Barcelo 71: 51–88. Elsevier.

Shelley, J. T., J. S. Wiley, and G. M. Hieftje. 2011. "Ultrasensitive Ambient Mass Spectrometric Analysis with a Pin-to-Capillary Flowing Atmospheric-Pressure Afterglow Source." *Analytical Chemistry* 83 (14): 5741–48.

Shiea, J., M. Z. Huang, H. J. Hsu, C. Y. Lee, C. H. Yuan, I. Beech, and J. Sunner. 2005. "Electrospray-Assisted Laser Desorption/Ionization Mass Spectrometry for Direct Ambient Analysis of Solids." *Rapid Communications in Mass Spectrometry* 19 (24): 3701–704.

Snyder, D. T., C. J. Pulliam, Z. Ouyang, and R. G. Cooks. 2016. "Miniature and Fieldable Mass Spectrometers: Recent Advances." *Analytical Chemistry* 88 (1): 2–29.

Song, Y., and R. G. Cooks. 2006. "Atmospheric Pressure Ion/Molecule Reactions for the Selective Detection of Nitroaromatic Explosives Using Acetonitrile and Air as Reagents." *Rapid Communications in Mass Spectrometry* 20: 3130–38.

Stoffels, A., J. Flikweert, W. W. Stoffels, and G. M. Kroesen. 2002. "Plasma Needle: A Non-destructive Atmospheric Plasma Source for Fine Surface Treatment of (Bio) Materials." *Plasma Sources Science and Technology* 11: 383–88.

Stopka, S. A., B. Shrestha, E. Marechal, D. Falconet, and A. Vertes. 2014. "Metabolic Transformation of Microalgae due to Light Acclimation and Genetic Modi Fi Cations Followed by Laser Ablation Electrospray Ionization Mass Spectrometry with Ion Mobility Separation." *Analyst* 139: 5945–53.

Sun, J., W. Wang, L. Xu, J. Dong, W. Gao, Z. Huang, P. Cheng, and Z. Zhou. 2015. "Development of an Air-Flow-Assisted Extractive Electrospray Ionization Source for Rapid Analysis of Phthalic Acid Esters in Spirits." *Rapid Communications in Mass Spectrometry* 29 (19): 1711–16.

Suni, N. M., H. Aalto, T. J. Kauppila, T. Kotiaho, and R. Kostiainen. 2012. "Analysis of Lipids with Desorption Atmospheric Pressure Photoionization-Mass Spectrometry (DAPPI-MS) and Desorption Electrospray Ionization-Mass Spectrometry (DESI-MS)." *Journal of Mass Spectrometry* 47 (5): 611–19.

Takáts, Z., J. M. Wiseman, B. Gologan, and R. G. Cooks. 2004. "Mass Spectrometry Sampling Under Ambient Conditions with Desorption Electrospray Ionization." *American Association for the Advancement of Science* 306 (5695): 471–73.

Tian, Y., J. Chen, Y. Z. Ouyang, G. B. Qu, A. F. Liu, X. M. Wang, C. X. Liu, J. B. Shi, H. W. Chen, and G. B. Jiang. 2014. "Reactive Extractive Electrospray Ionization Tandem Mass Spectrometry for Sensitive Detection of Tetrabromobisphenol A Derivatives." *Analytica Chimica Acta* 814: 49–54.

Vaclavik, L., T. Cajka, V. Hrbek, and J. Hajslova. 2009. "Ambient Mass Spectrometry Employing Direct Analysis in Real Time (DART) Ion Source for Olive Oil Quality and Authenticity Assessment." *Analytica Chimica Acta* 645: 56–63.

Venter, A. R., K. A. Douglass, J. T. Shelley, G. Hasman, and E. Honarvar. 2014. "Mechanisms of Real-Time, Proximal Sample Processing during Ambient Ionization Mass Spectrometry." *Analytical Chemistry* 86 (1): 233–49.

Verplanken, K., S. Stead, R. Jandova, C. V. Poucke, J. Claereboudt, J. V. Bussche, S. D. Saeger, Z. Takats, Jella Wauters, and Lynn Vanhaecke. 2017. "Rapid Evaporative Ionization Mass Spectrometry for High-Throughput Screening in Food Analysis: The Case of Boar Taint." *Talanta* 169: 30–36.

Wang, H., J. Liu, R. G. Cooks, and Z. Ouyang. 2010. "Paper Spray for Direct Analysis of Complex Mixtures Using Mass Spectrometry." *Angewandte Chemie - International Edition* 49 (5): 877–80.

Wang, N. N., H. D. Wang, J. H. Ding, Y. Z. Ouyang, X. B. Zhu, and H. W. Chen. 2014. "Detection of Plasticizers in Soil Using Surface Desorption Atmospheric Pressure Chemical Ionization Imaging Mass Spectrometry." *Chinese Journal of Analytical Chemistry* 42 (4): 547–51.

Wang, X., X. Li, Z. Li, Y. Zhang, Y. Bai, and H. Liu. 2014. "Online Coupling of In-Tube Solid-Phase Microextraction with Direct Analysis in Real Time Mass Spectrometry for Rapid Determination of Triazine Herbicides in Water Using Carbon-Nanotubes-Incorporated Polymer Monolith." *Analytical Chemistry* 86 (10): 4739–47.

Weston, D. J. 2010. "Ambient Ionization Mass Spectrometry: Current Understanding of Mechanistic Theory; Analytical Performance and Application Areas." *The Analyst* 135 (4): 661.

Wiley, J. S., Z. Takats, Z. Ouyang, and R. G. Cooks. 2015. "Desorption Electrospray Mass Spectrometry." In *Ambient Ionization Mass Spectrometry*, edited by M. Domin and R. Cody, 307–33. Cambridge, UK: Royal Society of Chemistry.

Wu, C., D. R. Ifa, N. E. Manicke, and R. G. Cooks. 2009. "Rapid, Direct Analysis of Cholesterol by Charge Labeling in Reactive Desorption Electrospray Ionization." *Analytical Chemistry* 81 (18): 7618–24.

Yang, J., Y. Li, Y. Wang, J. Ruan, J. Zhang, and C. Sun. 2015. "Recent Advances in Analysis of Phthalate Esters in Foods." *Trends in Analytical Chemistry* 72 (October): 10–26.

Yang, S., J. Ding, J. Zheng, B. Hu, J. Li, H. Chen, Z. Zhou, and X. Qiao. 2009. "Detection of Melamine in Milk Products by Surface Desorption Atmospheric Pressure Chemical Ionization Mass Spectrometry." *Analytical Chemistry* 81: 2426–36.

Zhang, H., A. Bibi, H. Lu, J. Han, and H. Chen. 2017. "Comparative Study on Ambient Ionization Methods for Direct Analysis of Navel Orange Tissues by Mass Spectrometry." *Journal of Mass Spectrometry* 52 (8): 526–33.

Zhang, L., W. Yong, J. Liu, S. Wang, Q. Chen, T. Guo, J. Zhang, T. Tan, H. Su, and Y. Dong. 2015. "Determination of Dicyandiamide in Powdered Milk Using Direct Analysis in Real Time Quadrupole Time-of-Flight Tandem Mass Spectrometry." *Journal of the American Society for Mass Spectrometry* 26 (8): 1414–22.

Zhang, Y., and H. K. Lee. 2013. "Liquid – Liquid Microextraction Combined with Gas Chromatography – Mass Spectrometry for the Fast Determination of Phthalate Esters in Bottled Water." *Journal of Chromatography A* 1274: 28–35.

Zhang, Z., R. G. Cooks, and Z. Ouyang. 2012. "Paper Spray: A Simple and Efficient Means of Analysis of Different Contaminants in Foodstuffs." *The Analyst* 137 (11): 2556.

Zhou, J., S. Yao, R. Qian, Z. Xu, Y. Wei, and Y. Guo. 2008. "Observation of Allicin-Cysteine Complex by Reactive Desorption Electrospray Ionization Mass Spectrometry for Garlic." *Rapid Communications in Mass Spectrometry* 22 (20): 3334–3337.

Zhou, S., M. W. Forbes, and J. P. D. Abbatt. 2015. "Application of Direct Analysis in Real Time-Mass Spectrometry (DART-MS) to the Study of Gas-Surface Heterogeneous Reactions: Focus on Ozone and PAHs." *Analytical Chemistry* 87 (9): 4733–40.

CHAPTER 11

Use of an "Intelligent Knife" (iknife), Based on the Rapid Evaporative Ionization Mass Spectrometry Technology, for Authenticity Assessment of Pistachio Samples*

Francesca Rigano, Sara Stead, Domenica Mangraviti, Renata Jandova, Davy Petit, Nino Marino, and Luigi Mondello

CONTENTS

* This chapter was originally published as an Open Access article in *Food Analytical Methods* (2019) 12:558–568. doi.org/10.1007/s12161-018-1386-8.

11.1 INTRODUCTION

The topic of food authenticity and genuineness centralizes special interest for reasons related to both economical and sociopolitical factors. From an economical point of view, it is essential to preserve the national producer, but at the same time, to create a competitive, sustainable, and innovative global market. On the other hand, the first important social aspect is the preservation of the consumers, by the provision of safe, nutritious, and high-quality food. Despite food labeling (Regulation EU 1169/ 2011), which aims to inform consumers about what kind of products they are buying and facilitate trade inside and outside Europe, the broad question of "food fraud" is not regulated by common EU legislation. The lack of such a mechanism often leads to the occurrence of a certain number of cross-border food law violations due to the difficulty for member states' competent authorities to communicate efficiently with their counterparts in other member states. In this framework, the so-called "General Food Law" (Regulation EC 178/ 2002) set out an overarching and coherent framework for the development of food and feed legislation both at union and national levels. In particular, the European Commission decided to activate a dedicated network of administrative assistance liaison bodies that would handle specific requests for cross-border cooperation. The dedicated liaison bodies are referred to as "Food Fraud Contact Points" (FFCP). The group of FFCPs is collectively referred to as the "Food Fraud Network" or FFN and aims to activate competent authorities against "food crime," promoting investigation/prosecution.

Within the network system, the EU Commission Joint Research Centre (JRC) ensures the maximum level of expertise and scientific knowledge in order to apply the best available science. In Italy, the JRC based in Ispra devoted special attention to the topic of food safety and security, aiming to preserve the vast national heritage of typical and unique foods, viz. the Protected Designation of Origin (PDO), Protected Geographical Indication (PGI), and Denomination of Controlled and Guaranteed Origin (DOCG) trademarks. This initiative calls for the development of advanced and innovative analytical technologies and methods, characterized by high sensitivity, selectivity, throughput, and feasibility. During the last decades, the exponential progress in capillary gas chromatography (GC) techniques and mass spectrometry (MS) instrumentation made GC and GC–MS the most used tool to detect adulteration or any kind of contamination (plastic compounds, pesticides) in food (Cubero-Leon et al. 2014). Afterward, in the last number of years, improvements in liquid chromatography (LC) columns and instrumentation and progress in high-resolution MS (HR-MS) incentivized the rapid development of metabolomics mainly based on the analysis of LC-amenable polar and/or large molecule (Wolfender et al. 2013) as food authenticity markers (polyphenols, amino acids, carotenoids, lipid species). For the sake of clarity, the main analytical techniques applied in this field are reported in Table 11.1 along with their benefits and drawbacks.

Two complementary metabolomics approach can be clearly identified: the first is based on the identification/quantification of one or more group of analytes usually by LC–MS techniques; the second one aims not to identify/quantify each observed molecular species, but to compare a fingerprint generated by the analysis of the whole metabolome. Shotgun

TABLE 11.1 Most Employed Analytical Techniques in the Field of Food Authenticity and Adulteration

Analytical Technique	Benefits	Drawbacks
GC-based approaches	Very high sensitivity	Limited to volatile and semi-volatile components
	High identification power	Need for derivatization or extraction procedures
LC–MS	Suitable for very large compounds (biomolecules)	Poor identification power compared to GC-EI-MS-based techniques
	High sensitivity	
Direct infusion MS	Very short analysis time	Low sensitivity for trace compounds
		Ion suppression/enhancement effects
		Unreliable quantification
Vibrational spectroscopy	Rapid and non-destructive analysis	Not suitable for high-moisture samples
	Very short analysis time	
NMR	High structural characterization power	Low sensitivity
	Very short analysis time	High maintenance cost
Bulk-IRMS	Direct origin/isotope ratio correlation	High probability of artifacts due to external contamination
		Difficult to detect adulteration at trace levels
GC-isotopic ratio mass spectrometry (IRMS)	Direct origin/isotope ratio correlation	Limited to volatile and semi-volatile components
		Need for pure compounds (baseline separation, no co-elution)
Sensory analysis	Low cost	Limited to a restricted number of molecules
Physicochemical characterization	Easy to use	Need for complementary analytical techniques

MS, vibrational spectroscopy (IR, near-IR (NIR), Raman, and FT Raman), and nuclear magnetic resonance (NMR) fall into the second approach, and take great advantage of the very short analysis time because of the absence of a chromatographic separation step and, sometimes, of complex sample preparation procedures. Furthermore, MS, especially HR–MS, being characterized by a high identification power, could be employed for the characterization of marker species.

Within this context, a very innovative analytical technique, namely iknife, based on a shotgun MS approach, could be a valid aid to obtain a fingerprint of different food products. The iknife, which means "intelligent knife," is a term coined in 2013 by Prof. Takats and co-workers (Balog et al. 2013) to describe the coupling between the electrosurgical tool (the knife) and an MS system based on a rapid evaporative ionization mechanism (rapid evaporative ionization mass spectrometry (REIMS)). Such a technique demonstrated significant potentiality in the clinical medical field, where a certain number of papers about cancer tissue identification has been published (Neidert and Bozinov 2013; Balog et al. 2013; Balog et al. 2015; Alexander et al. 2015; St John et al. 2017), by exploiting the near 100% accuracy in the in vivo recognition of cancer cells, in real time (3–4 s analysis time) without any sample preparation procedure. On the other hand, to the best of the authors' knowledge, only three papers refer to the use of the iknife or REIMS technology for the analysis of food, and all focus on animal tissues to discriminate different kinds of meat (Balog et al. 2016) or fish (Black et al. 2017) or identify boar taint in pig fat (Verplanken et al. 2017). Certainly, Italy may be the ideal location for the extension of such analytical technique in the food area, taking into account the significant number of PDO, PGI, and DOCG trademarks present in the territory. In this work, the novel iknife technique was exploited to obtain a clustering of various pistachio samples coming from different geographical origins. In particular, the main object was the preservation of the Bronte pistachio, a PDO trademark (OJEU 2009/C 130/09) that represents one of the greatest economical sources of the east Sicily area; it is cultivated around the Etna volcano in the area of Bronte, where the volcanic soil and climate allow the production of nuts with intense green color and aromatic taste, very appreciated in international markets. Due to the relatively small dimensions of the cultivated area, the Italian production is very low compared to Asia and California; nevertheless, the higher quality is recognized around the world. A further purpose of the present work was to determine marker compounds, responsible for the geographical differentiation. To confirm peak identification, conventional chromatographic techniques (GC–MS, LC–MS) were employed and compared to the novel iknife technique. The obtained results demonstrated the capability of the new technology to provide a highly informative holistic profile of a complex matrix.

11.2 MATERIAL AND METHODS

11.2.1 Reagents

LC–MS grade acetonitrile, methanol, water, and 2-propanol were purchased from Biosolve Chimie (Dieuze, France). Potassium hydroxide, n-hexane reagent grade, and 2-propanol HPLC grade were purchased from Carlo Erba (Milan, Italy). Ammonium formate was obtained from Alfa Aesar GmbH & Co KG (Karlsruhe, Germany). Standards of trinonanoin ($C_9C_9C_9$), triundecanoin ($C_{11}C_{11}C_{11}$), tritridecanoin ($C_{13}C_{13}C_{13}$), tripentadecanoin ($C_{15}C_{15}C_{15}$), triheptadecanoin ($C_{17}C_{17}C_{17}$), trinonadecanoin ($C_{19}C_{19}C_{19}$), and a C4–C24 even saturated fatty acid methyl ester (FAME) 1000 mg/L hexane solution were purchased from MilliporeSigma (Bellefonte, PA, USA).

11.2.2 Samples

Samples of unprocessed whole pistachio nuts from different geographical origins and varieties (Bronte, California prime quality, California second quality, Iran normal, Iran perfect green, Turkey mawardi, Turkey red, Turkey perfect green, Greece) were kindly provided by Pistì S.r.L. (Italy, Bronte, CT). A minimum of five production lots were collected for each variety.

11.2.3 Sample Preparation for REIMS Analysis

The pistachio nut outer shell and skin were removed by hand. Equal portions of the different production lots were combined and the nuts were subsequently crushed using a domestic coffee grinder device. A conductive paste was prepared via the addition of 200 μL deionized water per 1 g of crushed nut material. Approximately 1 g of the pistachio paste was placed on a wet tissue on top of the return electrode plate prior to iknife sampling.

11.2.4 REIMS Analysis

The iknife hand-held sampling device (Waters Corporation, Wilmslow, UK) was used to apply a localized high- frequency electric current to the surface of each sample, which instantly vaporizes molecules from the latter. It consisted of a monopolar cutting device with a shortened knife blade approximately 6 mm long and was applied in auto-cut mode in combination with a diathermy electrosurgical generator (Erbe VIO 50 C) (Erbe, Tuebingen, Germany) at a power of 30 W. Sampling was carried out for 3 to 5 s and, for each sample, technical replicates were analyzed, thus taking into account repeatability of the analysis. Mass spectrometric analysis was performed using a Xevo G2 XS QTOF instrument equipped with an REIMS source comprising a helical coiled ribbon collision surface heated by a constant current power supply set to 4.5 A and 4.2 V (Kanthal D 1.0 × 0.1 mm) (Waters Corporation, Wilmslow, UK). All analysis was performed in REIMS TOF MS sensitivity mode with continuum data acquisition. A matrix of 2-propanol was infused directly into the REIMS source at a constant flow rate of 200 μL min^{-1} to promote the ionization of lipid species and maintain source cleanliness. The mass resolution was approximately 20,000 FWHM over the mass range of interest. The cone voltage was set at 100 V. MS analysis was performed in negative ionization mode over a mass range of 50–1200 *m/z* with a (scan) acquisition time of 1 s/scan. Prior to use, the instrument was calibrated using sodium formate in 2-propanol and was infused via the matrix inlet on the REIMS source.

For quality control purposes, the endogenous matrix ion *m/z* 255.2330, corresponding to the deprotonated molecule of palmitic acid ($C_{16}H_{31}O_2$), was used for internal lock-mass correction. At least 20 replicate measurements were collected for each sample in two different laboratories (Waters-Wilmslow and Chromaleont-Messina) in order to obtain some inter-laboratory repeatability data and explore the possibility of transferring the statistical model between different geographical locations. Furthermore, replicate burns of a QC sample (porcine liver) were collected at the start and end of the analytical batch. The intensity of the base peak ion at *m/z* 699.497 was recorded and plotted for quality control monitoring. The iknife blade was cleaned using methanol after every 15 samples, and the 3-m-long transfer tubing and venturi air pump were cleaned in an ultrasonic bath using methanol at the end of each day.

An untargeted MS analysis was performed to discriminate between pistachio nuts grown in different geographical origins (Turkey, Greece, Iran, Sicily, and USA). In the case of pistachio nuts from the Turkey and Iran samples, more than one variety was present. The training samples were analyzed in two different laboratory locations on different days to take into account technical variation originating from intra- and inter-laboratory reproducibility. This step thus generated a total of 360 mass spectra (64 spectra for Sicilian Bronte 32 for Greek normal, 84 for Iranian normal and perfect green, 64 for USA prime and second quality, and 116 for Turkish mawardi, red, and perfect green) that were used to train the database. In addition, samples from a different production lot were collected and analyzed in triplicate and used as an independent validation challenge set.

The mass spectrometer was operated in REIMS MS/MS acquisition mode to generate CID fragments for ions identified as discriminatory features during the multivariate statistical analysis. Precursor ions were isolated in the resolving quadrupole region of the instrument. Argon was used as the collision gas, and collision energy was optimized on a compound by compound basis. The following MS/MS specific settings were applied, LM and HR resolution 10 and 14.5, respectively, prefilter 2.0 V, and ion energy 0.8 V.

11.2.5 Chemometric Data Analysis

Multivariate statistical software package LiveID™ (Waters Corporation, Wilmslow, UK) was used as a model builder and recognition tool. In order to generate models from the untargeted profiling REIMS TOF MS data acquired in MassLynx v. 4.1 (Waters Corporation, Wilmslow, UK), the following data treatment steps were performed: lock-mass correction applied using the endogenous matrix ion at *m/z* 255.2330; all spectra contained within each "burn event" termed the region of interest (ROI) were combined to form a single continuum spectrum; adaptive background subtraction (ABS) algorithm was applied to reduce the chemical background in the combined spectra; data resampling (binning to 0.1 Da) was performed to reduce the data dimensionality; and the resulting spectrum was normalized using the TIC. For principal component analysis (PCA), the data were centered using the mean value of the entire dataset. For linear discriminate analysis (LDA), the data were centered using the mean values of each model class. In either type, the mean for each *m/z* bin is subtracted from the values of that bin. Other than normalization and centering, no additional manipulation was performed (for example, scaling).

Following data pretreatment steps, a PCA/LDA model was calculated. Firstly, an unsupervised PCA (singular-value decomposition algorithm) transform is applied to the spectral data, calculating the scores and loadings; a supervised LDA transform is then applied to the scores calculated by the PCA transform. LDA is a transform that maximizes the inter-class variance, while minimizing the intra-class variance, resulting in a projection where examples from the same class are projected close to each other and, at the same time, the class centers (means) are as far apart as possible. Although it is not a true regularization technique, PCA-LDA is found to reduce the chance of overfitting that may occur with a pure LDA model.

During the recognition step, the model transforms spectra acquired from test samples into the associated model space, after which, a classifier decides to which class (if any) the spectra belong. The model classifier uses a multivariate normal distribution (MVN) for each model class. The MVN produce a likelihood measure for each class, and Bayes' rule was then applied to derive posterior probabilities.

In silico, fivefold stratified validation was performed to determine the predictive accuracy of the model. The model- building dataset was divided in five partitions (fivefold), each of which contains a representative proportion of each class within it (stratified). Four partitions (80%) of the dataset are used to build a model under the same conditions as the original model. This model is used to predict the classifications of the one partition (20%) of the training set that was left out. The cycle was repeated iteratively five times and each partition was predicted once by a model trained from the other four. The output of the validation details the total number of correct and incorrect classifications, as well as the number of outliers. Outliers were calculated according to the Mahalanobis distance to the nearest class center (Mahalanobis 1936).

If this distance is greater than the outlier threshold, the sample is considered an outlier. Following iterations of model optimization, an independent validation step was performed using a sample series not included in the training set.

Additional and complementary statistical analyses were performed using MetaboAnalyst 3.0 (Xia et al. 2015; Xia and Wishart 2016), Progenesis QI (Non-linear Dynamics, Newcastle, UK), EZInfo, and SIMCA-P (Umetrics Sartorius Stedim Biotech, Sweden) to determine candidate biomarkers and tentative assignments for preliminary pathway analysis.

11.2.6 Lipid Extraction for Chromatographic Analyses

Crushed Bronte pistachio nuts (2 g) underwent lipid extraction by using 6 mL of a hexane/2-propanol 3:2 (*v:v*) mixture at room temperature under vigorous stirring for 1 h. The homogenate was filtered and the residue was washed twice with 4 mL of the same extraction mixture. The filtrate was dried by using a rotary evaporator.

11.2.7 GC–MS Analysis of Total Fatty Acid Composition of Bronte Pistachio

The total fatty acid composition was investigated after the conversion of intact lipids into fatty acid methyl esters (FAMEs), by methanolic solution of potassium hydroxide at room temperature: 1 mL of hexane and 0.1 mL of methanolic potassium hydroxide solution 2 N were added to 50 mg of the lipid extract; the mixture was shaken for 30 s, and the upper hexane phase was injected into a GCMS-QP2010 (Shimadzu, Milan, Italy) equipped with a split-splitless injector and an AOC-20i auto-sampler. Separation was achieved by using a SLB-Il60i (30 m × 0.25 mm id. 0.20-µm film thickness) column (MilliporeSigma) under the following temperature program: from 50 to 280°C at 3.0°C/min. Injector was kept at 280°C; injection volume was 0.2 µL (split 1:100). Helium was used as the carrier gas at 30 cm/s linear velocity and a pressure of 27.7 KPa. MS parameters were as follows: mass range 40–550 amu with a scan interval of 0.20 s; ion source and interface temperatures were 200°C and 220°C, respectively. The GCMSsolution software version (Shimadzu) was used for data collection and handling. The C4–C24 even carbon saturated FAME standard solution was used for linear retention index (LRI) calculation to support identification. Then, peak assignment was based on a double filter (MS similarity over 85% and LRI ± 10) compared to lipids library (Shimadzu Europe, Duisburg, Germany).

11.2.8 LC–MS Analysis of Non-Polar Lipids

The lipid extract was injected into a Nexera X2 system coupled to a LCMS-2020 mass spectrometer (Shimadzu, Kyoto, Japan). Separation was carried out on a Titan C18 100 × 2 .1 mm (L × I D), 1 .9 µ m d p column (MilliporeSigma), operated at a 400 µL/min flow rate, by using acetonitrile (a) and 2-propanol (b) as mobile phase under gradient conditions: 0–52.5 min, 0–50% B. Oven temperature was set at 35°C and the injection volume was 2 µL. MS detection was performed by using APCI interface in positive ionization mode under the following conditions: the mass range was 300–1250 *m/z*, with an event time of 0.5 s; desolvation line, heat block, and interface temperatures were kept at 300°C,

300°C, and 350°C, respectively; nebulizer and drying gas (N_2) flows were set at 3 L/min and 5 L/min, respectively.

The recently introduced LRI filter was used as an additional identification tool, by using a $C_9C_9C_9$–$C_{19}C_{19}C_{19}$ odd carbon saturated triacylglycerol (TG) standard solution (1000 mg/L) as reference homolog series (Rigano et al. 2018). The analyses were acquired by using LabSolutions software version 5.91 (Shimadzu) and processed by using the Cromatoplus Spectra software (Chromaleont, Messina, Italy).

11.2.9 HILIC–MS Analysis of Polar Lipids

The lipid extract was injected into a Nexera X2 system coupled to an LCMS-2020 mass spectrometer (Shimadzu, Kyoto, Japan). Separation was carried out on an Ascentis Express HILIC 150 × 2.1 mm (L × ID), 2.7 µm dp column (MilliporeSigma), operated at a 300 µL/min flow rate, by using acetonitrile:10 mM ammonium formate 95:5 *v:v* (a) and acetonitrile:methanol:10 mM ammonium formate 55:35:10 *v:v:v* (b) as mobile phase under gradient conditions: 0–5 min, 0% B, 5–25 min, 0–40% B, 25–50 min, and 50–70% B. Oven temperature was set at 30°C and the injection volume was 5 µL. MS detection was performed by using ESI interface in both positive and ionization mode under the following conditions: the mass range was 200–1200 *m/z*, with an event time of 0.3 s; desolvation line and heat block temperatures were both kept at 230°C; nebulizer and drying gas (N_2) flows were set at 1.5 L/min and 5 L/min, respectively.

Data acquisition and processing were carried out by using LabSolutions software version 5.91 (Shimadzu).

11.3 RESULTS AND DISCUSSION

11.3.1 Mass Spectral Data

Pistachio nuts were widely studied in literature, mainly due to their beneficial effects on human health (Ryan et al. 2006; Tomaino et al. 2010; Mandalari et al. 2013; Chung et al. 2013; Song et al. 2018). Particularly, it is well-known that almost 50% of their chemical composition is constituted by lipids, essentially TGs, and it is well-known that the most abundant FAs are oleic, linoleic, and palmitic acid (Ryan et al. 2006; Chung et al. 2013; Song et al. 2018). Since the majority of applications of the REIMS technique deal with the untargeted profile of the lipid fraction, such as FAs and glycerophospholipids (PLs) in negative ionization mode (Schafer et al. 2009; Balog et al. 2010; Balog et al. 2013; Strittmatter et al. 2014; Alexander et al. 2015; Balog et al. 2015; Golf et al. 2015a; Golf et al. 2015b; Balog et al. 2016; Strittmatter et al. 2016; Black et al. 2017; St John et al. 2017; Verplanken et al. 2017), the comparison of the lipid profile between nine pistachio varieties was exploited also in this study for the authenticity assessment of the geographical origin. Figure 11.1a reports the typical total ion current (TIC) chromatogram generated by 20 subsequent cuts of a pistachio sample in order to insert in the statistical model a sufficient number of spectra replicates, if possible, acquired by different operators and on different crushed nuts. The average ion intensity was around 35 million for all cuts with a standard deviation of ±10 million, corresponding to a relative standard deviation percentage of less than 30%, thus revealing a satisfactory repeatability of manual cuts. The typical MS spectrum in negative ionization mode is shown in Figure 11.1b; the large mass range investigated (50–1200

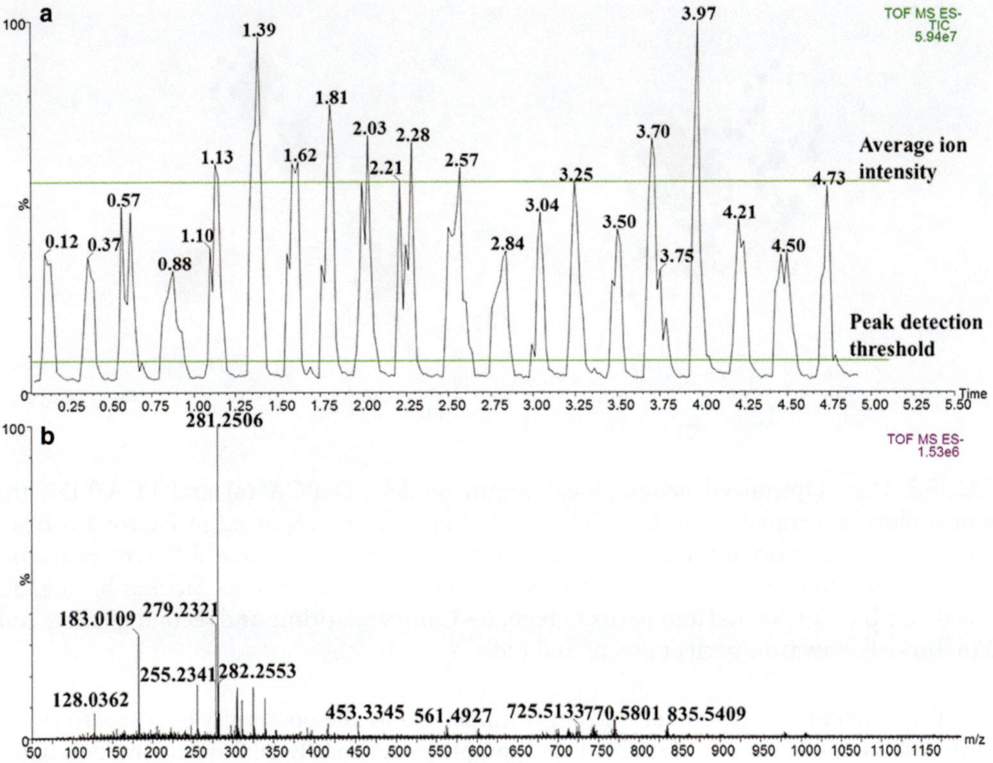

FIGURE 11.1 (a) Typical total ion count chromatogram showing 20 replicate iknife sampling events taken from Sicilian Bronte pistachio nut paste; (b) combined mass spectrum (ten scans), accurate mass corrected to the ion at 255.23 *m/z* and background subtracted, acquired using REIMS ionization in negative polarity over 50–1200 *m/z* range.

m/z) allowed the detection of not only FAs, PLs, and TGs, but also small molecules such as amino acids and phenolic acids. Actually, the mass range was slightly reduced to build the statistical model, since in silico validation test using the novel LiveID software demonstrated that the highest correctness score could be achieved by considering the mass range 200–1000 *m/z*, including both FAs and PLs, while excluding smaller molecules. Moreover, in order to improve spectra quality, a 6-million peak detection threshold was set.

11.3.2 Model Building and Validation 1: Classification According to the Geographical Origin

The results from the in silico stratified fivefold validation of the geographical origin of pistachio nut model, iteratively generated using different fitting parameters using the LiveID™ software, are reported in Table 1-S, while Figure 1-S contains the confusion matrix generated using the in silico validation tool in LiveID for the parameters selected as optimum. Only 1 out of 360 spectra was wrongly identified by the model, thus demonstrating the accuracy of approximately 100% in the identification of unknowns. Moreover the high identification power is ensured by the fact that at least three replicates of tentative identifications should be taken into account for the experimental validation of the model.

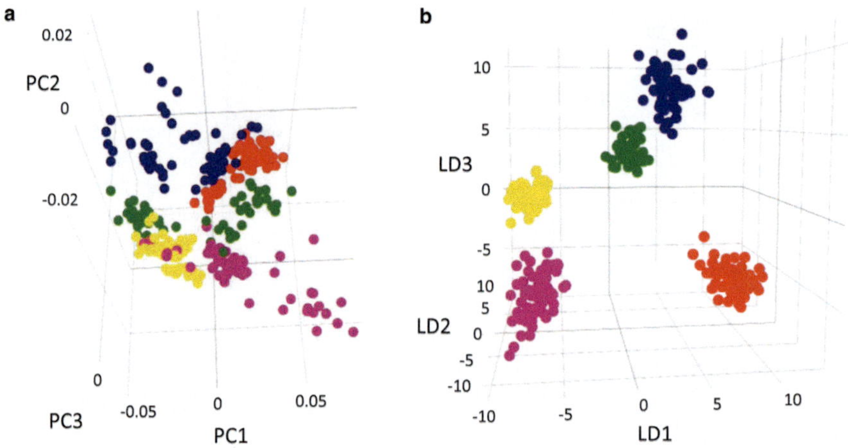

FIGURE 11.2 Optimized geographical origin model 3D PCA (a) and PCA/LDA (b) scores plots, generated using LiveIDTM, using parameters reported in Figure 1-S from REIMS negative polarity TOf-MS inter-laboratory data from three different pistachio lots, from different geographical origins for a total of 360 spectra (64 Sicilian Bronte, 32 Greek, 84 Iranian normal and perfect green, 64 California prime and second quality, and 116 Turkish mawardi, perfect green, and red).

The reliability of the model is strongly influenced by the possibility of access to really authentic samples, since a "false"-labeled sample could totally corrupt the model, hampering any accurate identification.

The tridimensional visualization of the validated model, both PCA and PCA/LDA score plots, is reported in Figure 11.2a, b. It enables only the classification/clusterization according to the geographical origin, not providing any information about quality (California prime and second quality) or maturation degree (different kind of Turkish and Iranian nuts). Instead, the analyses on all nine pistachio varieties were used to build a big five-class model, where each class corresponds to each origin.

Three production lots and analyses performed by different operators in different analytical laboratories were used for model building, thus performing for the first time since the introduction of this novel technique in the food field an inter-laboratory study, maximizing the analytical variability. The model was subsequently challenged using 12 different samples not included in the model training set originating from a different production lot, analyzed in triplicate using the real-time recognizer functionality. The overall outcome was determined from the mean of the three replicates (Table 2-S). Individual results having a % confidence factor of less than 80% were excluded from the overall outcome. On the basis of the independent validation, all samples were correctly classified leading to an overall correct assignment rate of 100%. Based on this validation set, a 0% false positive rate and false negative rate were estimated. Such preliminary performances are very satisfactory, especially considering that different varieties have been combined.

The same results can be highlighted from the playback recognition performed by the LiveID software for all the geographical origins (Figures 2-S, A–E).

Since the LiveID software is a novel software, developed ad hoc for REIMS applications, one of the most employed of the software for statistical data processing, namely MetaboAnalyst by means of OPLS-DA data analysis algorithm, was also used for comparison purposes. The principal component analysis and OrthoPLS-DA scores plot can be visualized in Figure 3-S A and B, showing already in the unsupervised analysis (Figure 3-S A) a clear trace of subdivision in five elliptical regions according to the origin. Such a result confirmed the ability of the studied ionization strategy to provide specific fingerprints of pistachio nuts in relation to geographical/climatic region.

11.3.3 Candidate Biomarkers for Bronte Pistachio Nuts

After model building, the corresponding loading plots for the main principal components were carefully investigated to reveal significant ions responsible for sample differentiation. In particular, Figure 11.3a, b shows the loading plots in the first two components, contributing to the discrimination with a percentage of about 57% and 17%, respectively. The base peaks in both cases correspond mainly to FA species, including palmitic and oleic acid (*m/z* 255.23 and *m/z* 281.25, respectively) in the positive direction of the first component, linoleic acid (*m/z* 279.23) in the negative direction of the second component.

As alternative/complementary software, Progenesis QI and EZInfo were used to show the significant features (*m/z*) responsible for the discrimination between Sicilian Bronte pistachio and all the other origins (Figure 4-S A–D). Such loading plots allowed selected biomarkers to be found for Bronte pistachio nuts, according to one of the main purposes of the present work to preserve the precious Sicilian variety. In order for an ion to be relevant a cutoff value of $p > 0.01$ (by ANOVA test, fold change ≥2) was established. Also, in this case, the novel LiveID software provided very similar results with respect to more consolidated software packages. The main fragment ions revealed from both software were selected as candidate biomarkers and are reported in Table 11.2.

MetaboAnalyst software was also employed to identify the features responsible for the discrimination between pistachio nuts from different geographical origins, obtaining many *m/z* values already highlighted by using both LiveID and Progenesis QI software. Figure 5-S represents a hierarchal cluster analysis heat map, generated using MetaboAnalyst 3.0 software, summarizing the features responsible for the discrimination between pistachio nuts from all five different geographical origins (ANOVA p value ≤ 0.05 max fold change ≥2 and minimum CV ≥ 30). All these compounds have been already assigned as candidate biomarkers after data processing by LiveID and Progenesis QI software (Table 11.2), thus representing a further confirmation of the proper applicability of the novel software to REIMS analyses.

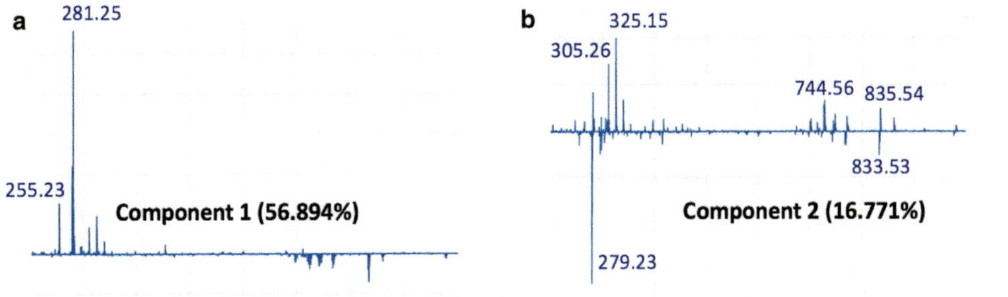

FIGURE 11.3 Loading plots generated by the LiveID software showing the *m/z* features responsible for the discrimination in the region between 200 and 1000 *m/z* in the first (a) and (b) in the second PCA component.

TABLE 11.2 Tentative Identifications for a Selection of the Discriminatory Features for Sicilian Bronte Pistachio Highlighted as Significant Following Multivariate Statistical Analysis (ANOVA $p > 0.01$, Fold Change ≥2)

Significant Ion (*m/z*)	Tentative Assignment	Main Class	Elemental Composition	Ion Detected
255.2328	Palmitic acid (C16:0)	Fatty acids and conjugates	C16H32O2	$[M–H]^-$
281.2488	Oleic acid (C18:1n9)	Fatty acids and conjugates	C18H34O2	$[M–H]^-$
283.2621	Stearic acid (C18:0)	Fatty acids and conjugates	C18H36O2	$[M–H]^-$
279.2321	Linoleic acid (C18:2n6)	Fatty acids and conjugates	C18H32O2	$[M–H]^-$
309.1729	Gadoleic acid (C20:1n9)	Fatty acids and conjugates	C20H34O2	$[M–H]^-$
367.3579	Lignoceric acid (C24:0)	Fatty acids and conjugates	C24H48O2	$[M–H]^-$
722.4880	PG (C16: 0/C16:0)	Phosphatidylglycerols	C38H74O10P	$[M–H]^-$
833.5264	PI (C16:0/C18:2)	Phosphatidylinositols	C43H79O13P	$[M–H]^-$
835.5418	PI (C16:0/C18:1)	Glycerophosphoinositols	C43H81O13P	$[M–H]^-$
881.7622	TG (C18:1/C18:1/C18:2)	Triacylglycerols	C57H102O6	$[M–H]^-$
883.7815	TG (C18:1/C18:1/C18:1)	Triacylglycerols	C57H104O6	$[M–H]^-$

The identification of PLs and TAGs was confirmed by MS/MS experiments. Particularly MS/MS spectra highlighted the right PL class, FA combination, and positional isomer. As an example, Figure 11.4 reports the MS/MS spectra of phospholipidinositol PI (C16:0/C18:1); the losses of inositol polar head, C18:1 and C16:0 FA, indicates the PI class and the single-FA combination, while fragment ratio and the presence of an intense *m/z* 391.2265 value, corresponding to the loss of inositol and C18:1, and lead to the correct positional isomer, where C18:1 occupies the sn1 PL position.

In addition, a conventional GC–MS analysis was performed for the elucidation of the total FA composition, particularly to confirm the presence of FAs normally not reported in previous pistachio works (Ryan et al. 2006; Chung et al. 2013), such as C24:0. The total FA profile is reported in Figure 11.5a, along with peak identification.

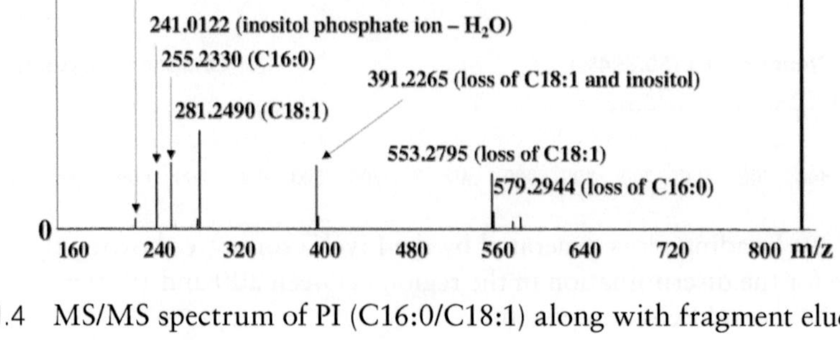

FIGURE 11.4 MS/MS spectrum of PI (C16:0/C18:1) along with fragment elucidation.

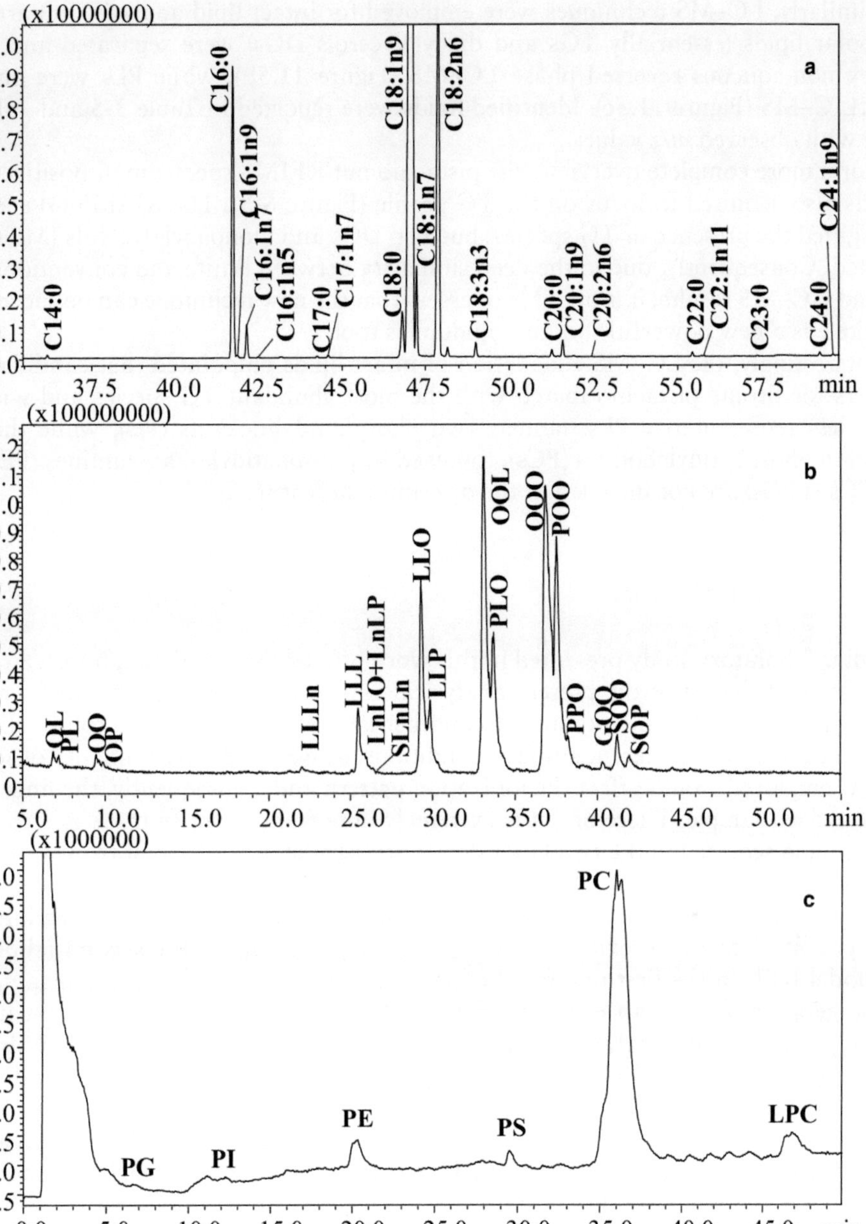

FIGURE 11.5 (a) GC–MS profile of FAMEs obtained from Bronte pistachio nuts; (b) reversed phase LC–MS profile of acylglycerols in the lipid extract of Bronte pistachio nuts; (c) HILIC–MS profile of PLs in the lipid extract of Bronte pistachio nuts. Fatty acid legend: P, palmitic acid C16:0; Ln, linolenic acid C18:3n3; L, linoleic acid C18:2n6; O, oleic acid C18:1n9; S, stearic acid; G, gadoleic acid C20:1n9; phospholipid legend; PG, phosphatidylglycerol; PI, phosphatidylinositol; PE, phosphatidylethanolamine; PC, phosphatidyl- choline; LPC, lysoPC.

Similarly, LC–MS techniques were employed for intact lipid analysis. In particular, non-polar lipids (essentially TGs and diacylglycerols DGs) were separated and identified by non-aqueous reversed phase LC–MS (Figure 11.5b), while PLs were analyzed by HILIC–MS (Figure 11.5c). Identified lipids were reported in Table 3-S and Table 4-S along with observed *m/z* values.

For a more complete overview, the pistachio nut REIMS spectrum in positive polarity was also acquired to focus on the TG profile (Figure 6-S). The REIMS (+) spectrum highlighted the presence of TG species, but also DGs and monoacylglycerols (MGs) were detected. Consequently, due to the deep similarity between iknife and conventional LC–MS and GC–MS results, it is possible to assess that the new technique can be successfully exploited as a new powerful shotgun lipidomics tool.

Interestingly, the LC–MS elucidation of intact lipids pinpointed that candidate biomarkers for Bronte pistachio match with the most abundant TG species and with two of the less representative PLs, namely two phosphatidylinositols (PIs), while the most abundant phosphatidylcholines (PCs), followed by phosphatidylethanolamines (PEs), and LysoPCs (LPCs) are not included in the discriminant features.

11.3.4 Model Transferability

The inter-laboratory study presented in this work offered the possibility, for the first time in the state of the art of the present analytical approach, to transfer the model entirely built in a given analytical laboratory to another laboratory, located in a totally different geographical region. Different ambient conditions, as well as different quality of solvents and gas employed, could affect the ionization pattern and, consequently, the fingerprinting of a given sample. Therefore, it is extremely important to verify if the spectra of the same sample acquired in the two laboratories are significantly different from a statistical standpoint.

This point was assessed not only by validating a model including both laboratories' spectral data (see previous section), but also by using, in the second laboratory, the model built in the first one for the identification of a set of validation samples of all the geographies. As an example, Figure 11.6 reports the correct identification of Bronte pistachio nuts analyzed in Messina, according to a model built in Wilmslow. The recognition occurs for more subsequent cuts with an accuracy of around 100%, by setting a maximum standard deviation of 30 units, meaning that two samples are recognized as "identical" if they differ by less than this threshold (in this case the samples are not statistically different). In effect, the spectra comparison reported in Figure 7-S A and B highlights slight differences in the profiling, but this is not the main object of this novel approach, aiming to create a database of samples, rather than a spectral library.

The model transferability here demonstrated represents a valid and powerful starting point for the establishment of the novel technique in the food application field, since it should be possible to build a model in an analytical laboratory and use it in another laboratory for authenticity evaluation and for the safeguarding of PDO, PGI, or DOCG products. In other words, a research laboratory can build a suitable model for discriminating a specific product, and the same model could be employed by the cross-border investigative authorities to detect fraudulent practices.

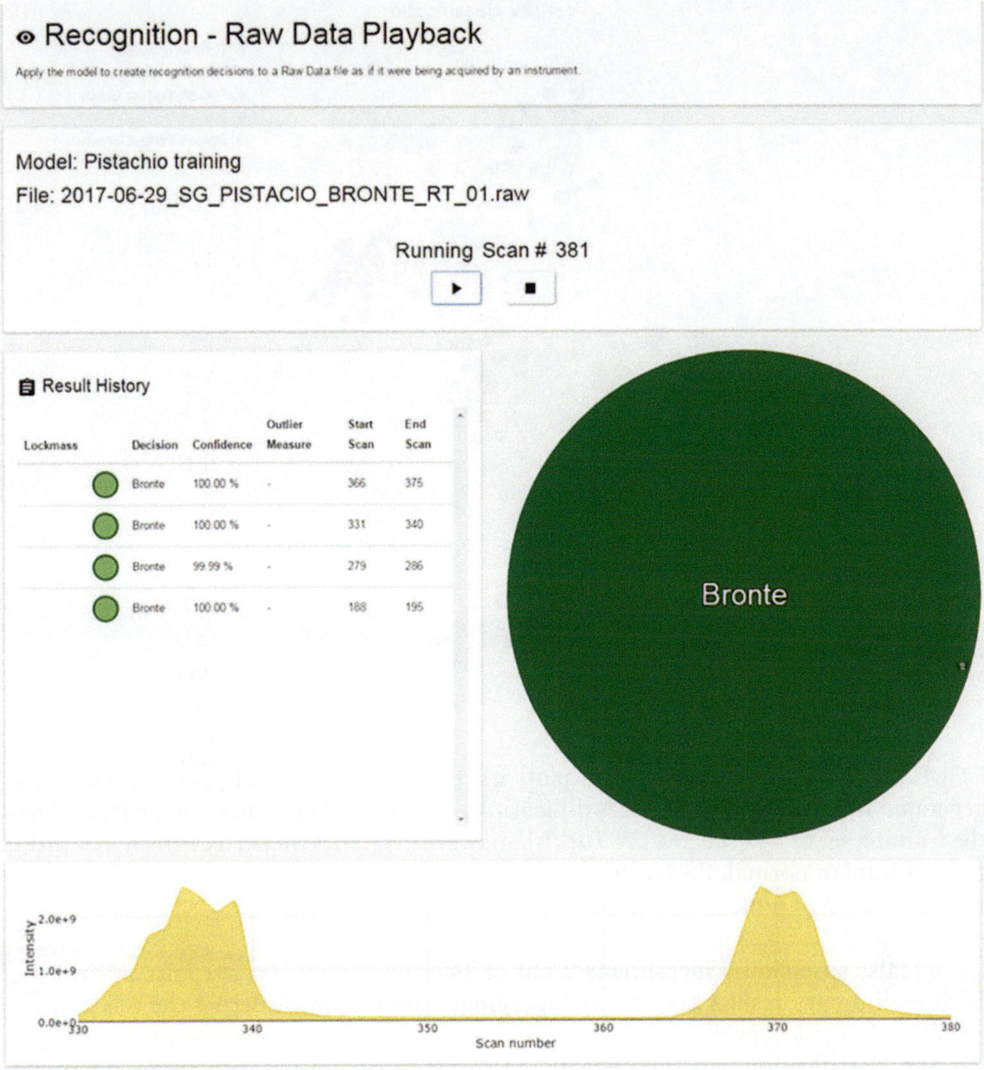

FIGURE 11.6 Identification by the LiveID software of Sicilian Bronte pistachio nuts analyzed in Messina, according to a model built in Wilmslow.

11.3.5 Model Building and Validation 2: Classification According to the Variety

A two-tier classification system was devised whereby a second model was trained to recognize the specific variety in the case of the Iranian and Turkish pistachio nuts following the initial classification of geographical origin (Figure 11.7). The same iterative approach described above was used for the second model optimization. The following parameters were selected for the variety classification model: *m/z* 200–1000, 20 PCA dimensions followed by four LDA dimensions. The predictive classification accuracy was calculated around 98% via stratified fivefold in silico validation, which presented only

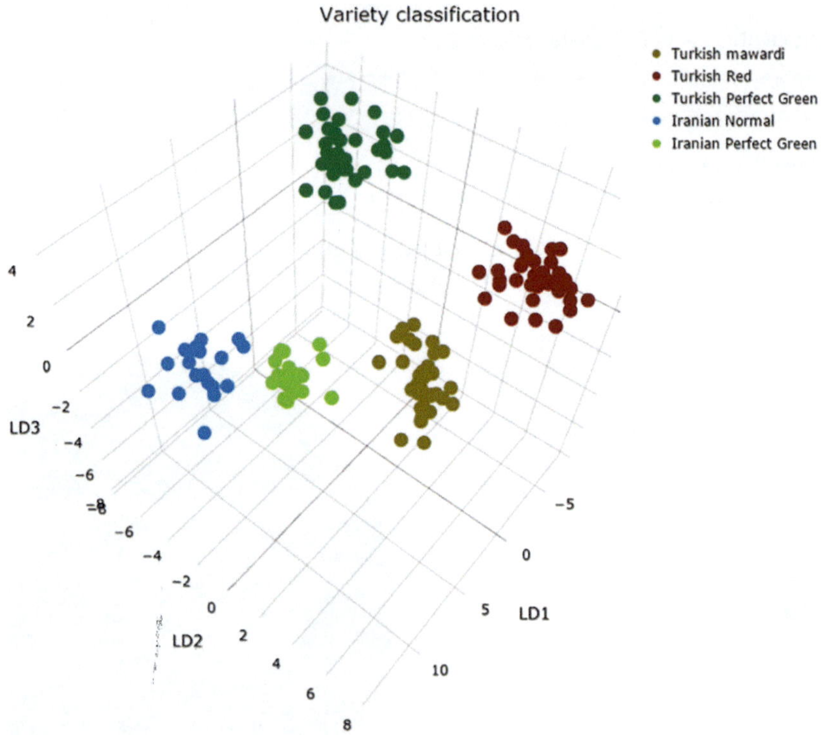

FIGURE 11.7 3D PCA/LDA scores plots for sub-classification of pistachio nut variety generated in LiveID using 20 PCA dimensions and four LDA dimension generated from the training set of 147 spectra (34 Turkish mawardi, 35 Turkish perfect green, 35 Turkish red, 20 Iranian normal, 23 Iranian perfect green) across a mass range of 200–1000 *m/z* with a binning parameter of 0.1 Da.

three "false suspicious" measurement out of 147 analyses, viz. a 2% false positive rate. Such preliminary results are very encouraging, especially considering the relatively low number of samples involved.

Sub-classification was subsequently performed using the variety model for samples classified as Iranian or Turkish.

Table 5-S reports the experimental validation results obtained on samples not employed for model building. A correct sample assignment (for three replicates) was achieved for all the varieties and production lots, in both analytical laboratories and considering analyses carried out by different operators. As couple of examples, Figure 8-S A and B provides the correct recognition of both Turkish and Iranian samples.

11.4 CONCLUSIONS

The technique described in the present research demonstrated very high potentiality in the differentiation of pistachio nuts coming from different geographical origins. The main goal with respect to other food applications of the iknife technique, where different animal species or genders have been successfully compared, was that exactly the same botanical species was analyzed for all the geographical origins. As a consequence, the present work could be the starting point of a more extensive project that aims to preserve the finest and most famous "Made in Italy" products.

Moreover, some discussions with small local producers suggest the creation og a similar model for pistachio-derived products, such as cream, ice-cream, and pesto, since the probability of detecting fraudulent actions significantly increases with respect to pure pistachio nuts. In addition, because of the high probability of finding mixtures of different pistachio in the final derived products, it will be very helpful to include in the database mixtures of pistachio nuts at known percentage ratios.

Moreover, an inter-laboratory study was performed for the first time, confirming the potentiality of the new technique to be easy to use by any operator, even not expert, after the building of a reliable database in an analytical laboratory.

Finally, the comparison with conventional chromatographic techniques made the identification of candidate biomarkers very reliable, unlike the tentative assignment of previous works.

COMPLIANCE WITH ETHICAL STANDARDS

Conflict of Interest Francesca Rigano declares that she has no conflict of interest. Sara Stead declares that she has no conflict of interest. Domenica Mangraviti declares that she has no conflict of interest. Renata Jandova declares that she has no conflict of interest. Davy Petit declares that he has no conflict of interest. Luigi Mondello declares that he has no conflict of interest.

Ethical Approval This article does not contain any studies with human participants or animals performed by any of the authors.

Informed Consent Informed consent is not applicable for this study.

REFERENCES

Alexander J, Gildea L, Balog J, Speller A, McKenzie J, Muirhead L, Scott A, Kontovounisios C, Rasheed S, Teare J, Hoare J (2015) A novel methodology for in vivo endoscopic phenotyping of colorectal cancer based on real-time analysis of the mucosal lipidome: a prospective observational study of the iKnife. *Surg Endosc* 31(3):1361–1370.

Balog J, Szaniszlo T, Schaefer K-C, Denes J, Lopata A, Godorhazy L, Szalay D, Balogh L, Sasi-Szabo L, Toth M, Takats Z (2010) Identification of biological tissues by rapid evaporative ionization mass spectrometry. *Anal Chem* 82:7343–7350.

Balog J, Sasi-Szabó L, Kinross J, Lewis MR, Muirhead LJ, Veselkov K, Mirnezami R, Dezső B, Damjanovich L, Darzi A, Nicholson JK, Takáts Z (2013) Intraoperative tissue identification using rapid evaporative ionization mass spectrometry. *Sci Transl Med* 5(194):194ra93.

Balog J, Kumar S, Alexander J, Golf O, Huang J, Wiggins T, Abbassi-Ghadi N, Enyedi A, Kacska S, Kinross J, Hanna GB, Nicholson JK, Takats Z (2015) In vivo endoscopic tissue identification by rapid evaporative ionization mass spectrometry (REIMS). *Angew Chem Int Ed* 54:11059–11062.

Balog J, Perenyi D, Guallar-Hoyas C, Egri A, Pringle SD, Stead S, Chevallier OP, Elliott CT, Takats Z (2016) Identification of the species of origin for meat products by rapid evaporative ionization mass spectrometry. *J Agric Food Chem* 64:4793–4800.

Black C, Chevallier OP, Haughey SA, Balog J, Stead S, Pringle SD, Riina MV, Martucci F, Acutis PL, Morris M, Nikolopoulos DS, Takats Z, Elliott CT (2017) A real time metabolomic profiling approach to detecting fish fraud using rapid evaporative ionisation mass spectrometry. *Metabolomics* 13:153–165.

Chung KH, Shin K, Hwang HJ, Choi K-S (2013) Chemical composition of nuts and seeds sold in Korea. *Nutr Res Pract* 7(2):82–88.

Cubero-Leon E, Peñalver R, Maquet A (2014) Review on metabolomics for food authentication. *Food Res Int* 60:95–107.

Golf O, Muirhead LJ, Speller A, Balog J, Abbassi-Ghadi N, Kumar S, Mróz A, Veselkov K, Takáts Z (2015a) XMS: cross-platform normalization method for multimodal mass spectrometric tissue profiling. *J Am Soc Mass Spectrom* 26:44–54.

Golf O, Strittmatter N, Karancsi T, Pringle SD, Speller AVM, Mroz A, Kinross JM, Abbassi-Ghadi N, Jones EA, Takats Z (2015b) Rapid evaporative ionization mass spectrometry imaging platform for direct mapping from bulk tissue and bacterial growth media. *Anal Chem* 87:2527–2534.

Mahalanobis PC (1936) On the generalised distance in statistics. *Proc Natl Inst Sci India* 12:49–55.

Mandalari G, Bisignano C, Filocamo A, Chessa S, Sarò M, Torre G, Faulks RM, Dugo P (2013) Bioaccessibility of pistachio polyphenols, xanthophylls, and tocopherols during simulated human digestion. *Nutrition* 29:338–344.

Neidert MC, Bozinov O (2013) Mass spectrometry-based intraoperative tissue identification in neurosurgery. *World Neurosurg* 80:683–684.

Official Journal European Union 2009/C 130/09: Publication of an application pursuant to Article 6(2) of Council Regulation (EC) No 510/ 2006 on the protection of geographical indications and designations of origin for agricultural products and foodstuffs.

Regulation (EC) 178/2002 of the European Parliament and of the Council of 28 January 2002.

Regulation (EU) 1169/2011 of the European Parliament and of the Council of 25 October 2011.

Rigano F, Oteri M, Russo M, Dugo P, Mondello L (2018) Proposal of a linear retention index system for improving identification reliability of triacylglycerol profiles in lipid samples by liquid chromatography methods. *Anal Chem* 90:3313–3320.

Ryan E, Galvin K, O'Connor TP, Maguire AR, O'Brien NM (2006) Fatty acid profile, tocopherol, squalene and phytosterol content of Brazil, pecan, pine, pistachio and cashew nuts. *Int J Food Sci Nutr* 57:19–228.

Schafer K-C, Denes J, Albrecht K, Szaniszló T, Balog J, Skoumal R, Katona M, Tóth M, Balogh L, Takáts Z (2009) In vivo, in situ tissue analysis using rapid evaporative ionization mass spectrometry. *Angew Chem Int Ed Engl* 48:8240–8242.

Song S, Cheong L-Z, Wang H, Man QQ, Pang SJ, Li YQ, Ren B, Wang Z, Zhang J (2018) Characterization of phospholipid profiles in six kinds of nut using HILIC-ESI-IT-TOF-MS system. *Food Chem* 240:1171–1178.

St John ER, Balog J, McKenzie JS, Rossi M, Covington A, Muirhead L, Bodai Z, Rosini F, Speller AV, Shousha S, Ramakrishnan R (2017) Rapid evaporative ionisation mass spectrometry of electrosurgical vapours for the identification of breast pathology: towards an intelligent knife for breast cancer surgery. *Breast Cancer Res* 19:59–72.

Strittmatter N, Rebec M, Jones EA, Golf O, Abdolrasouli A, Balog J, Behrends V, Veselkov KA, Takats Z (2014) Characterization and identification of clinically relevant microorganisms using rapid evaporative ionization mass spectrometry. *Anal Chem* 86:6555–6562.

Strittmatter N, Lovrics A, Sessler J, McKenzie JS, Bodai Z, Doria ML, Kucsma N, Szakacs G, Takats Z (2016) Shotgun lipidomic profiling of the NCI60 cell line panel using rapid evaporative ionization mass spectrometry. *Anal Chem* 88:7507–7514.

Tomaino A, Martorana M, Arcoraci T, Monteleone D, Giovinazzo C, Saija A (2010) Antioxidant activity and phenolic profile of pistachio (*Pistacia vera* L., variety Bronte) seeds and skins. *Biochimie* 92:1115–1122.

Verplanken K, Stead S, Jandova R, Poucke CV, Claereboudt J, Bussche JV, Saeger SD, Takats Z, Wauters J, Vanhaecke L (2017) Rapid evaporative ionization mass spectrometry for high-throughput screening in food analysis: the case of boar taint. *Talanta* 169:30–36.

Wolfender J-L, Rudaz S, Choi YH, Kim HK (2013) Plant metabolomics: from holistic data to relevant biomarkers. *Curr Med Chem* 20(8):1056–1090.

Xia J, Sinelnikov IV, Han B, Wishart David S (2015) MetaboAnalyst 3.0—making metabolomics more meaningful. *Nucl Acids Res* 43(W1):W251–W257.

Xia J, Wishart DS (2016) Using MetaboAnalyst 3.0 for comprehensive metabolomics data analysis. *Curr Protoc Bioinform* 55:14.10.1–14.10.91.

SECTION 4

Data Handling

CHAPTER 12

Comparison of Machine Learning Algorithms for Predictive Modeling of Beef Attributes Using Rapid Evaporative Ionization Mass Spectrometry (REIMS) Data*

Devin A. Gredell, Amelia R. Schroeder, Keith E. Belk, Corey D. Broeckling, Adam L. Heuberger, Soo-Young Kim, D. Andy King, Steven D. Shackelford, Julia L. Sharp, Tommy L. Wheeler, Dale R. Woerner, and Jessica E. Prenni

CONTENTS

Ambient mass spectrometry is an analytical approach that enables the ionization of molecules under open-air conditions with no sample preparation and very fast sampling times. Rapid evaporative ionization mass spectrometry (REIMS) is a relatively new type of ambient mass spectrometry that has demonstrated applications in both human health and food science. Here, we present an evaluation of REIMS as a tool to generate molecular-scale information as an objective measure for the assessment of beef quality attributes. Eight different machine-learning algorithms were compared to generate predictive models using REIMS data to classify beef quality attributes based on the United States

* This chapter was originally published as an Open Access article in *Scientific Reports* (2019), 9:5721. doi.org/10.1038/s41598-019-40927-6.

Department of Agriculture (USDA) quality grade, production background, breed type, and muscle tenderness. The results revealed that the optimal machine-learning algorithm, as assessed by predictive accuracy, was different depending on the classification problem, suggesting that a "one-size-fits-all" approach to developing predictive models from REIMS data is not appropriate. The highest performing models for each classification achieved prediction accuracies between 81.5 and 99%, indicating the potential of the approach to complement current methods for classifying quality attributes in beef.

In food science, chemical screening of ingredients and finished products is critical to ensure quality for food producers and consumers. Mass spectrometry (MS) is an important chemical detection platform for food analysis; however, methods typically require lengthy and complex sample preparation steps and analysis times.[1] Ambient mass spectrometry is a relatively new approach that enables the ionization of molecules under ambient conditions with no sample preparation and very fast sampling times. Takats et al. reported the first ambient ionization approach, desorption electrospray ionization (DESI), in 2004.[2] There are now over 30 reported ambient ionization techniques spanning application areas from pharmaceutical analysis to biological imaging to forensics and explosives detection.[3] The application of ambient ionization technology for the analysis of food and in particular for the detection of food fraud was recently reviewed by Black et al.[1]

Rapid evaporative ionization mass spectrometry (REIMS) is an emerging ambient ionization technique that has demonstrated applications in both human medicine and food science.[4] For example, REIMS-based tissue analysis can be used for intraoperative analysis of histological tissue for the identification of cancerous tissue margins and other prognostic and diagnostic applications.[5–7] While the REIMS technology was initially developed with biomedical applications in mind, it has also proven to be a valuable tool for the analysis of food. Recently, the utilization of REIMS for the analysis of meat products has generated very promising results across various classification scenarios reflective of important quality attributes such as genetic differences, production background, and sensory attributes. Balog et al. (2016) used REIMS to differentiate between various mammalian meat species and beef breeds with 100% and 97% accuracy, respectively.[8] Similarly, Black et al. used REIMS to accurately (98.9%) classify several fish species as an approach for detecting food fraud in the seafood industry.[9] Guitton et al. (2018) utilized REIMS to detect lipid changes in porcine muscle tissue reflective of treatment with ractopamine, a common growth-promotant used to increase muscle mass in swine.[10] Verplanken et al. (2017) successfully segregated pork carcasses with and without boar taint, an important sensory attribute related to pork quality.[11] Importantly, these examples illustrate the potential value of using REIMS to generate molecular-scale information as an objective measure for the assessment of meat quality.

To utilize molecular profiles generated by REIMS or any of the ambient ionization techniques as a means to classify samples, one must employ machine-learning algorithms to generate a predictive model. Machine learning is the process of rapidly finding and characterizing patterns in complex data.[12] There are many different types of machine learning including, for example, decision tree learning, network analysis, linear regression, support vector machines, and similarity functions.[12] Each of these algorithms is based on different mathematical approaches and thus, it is expected that with variation in the types of samples, data, and phenotypes in an experiment, one algorithm can greatly outperform others in terms of prediction accuracy. The majority of REIMS applications have used linear discriminant analysis on principal component analysis (PCA-LDA) reduced data for the generation of predictive models. The PCA-LDA method performs well for the classification of groups that tend to show large differences in the molecular profile of samples, such as the REIMS-meat studies described above. However, when molecular profiles of samples are not overly distinct (e.g. consumer food preference) or classification of multiple groups within a single model is desired, alternative machine-learning approaches may outperform PCA-LDA.[13]

In this study, the predictive accuracy of eight different machine-learning algorithms were compared for the generation of predictive models using REIMS data to classify attributes of beef based on USDA quality grade, production background, breed type, and muscle tenderness.

12.1 METHODS

Institutional Animal Care and Use Committee approval was not required for this study as samples were obtained postmortem from federally inspected harvest facilities.

Sample Collection. Beef strip loin sections (*longissimus* muscle) were collected to represent seven carcass types [Select (n = 42), Low Choice (n = 42), Top Choice (n = 39), Prime (n = 42), Dark Cutter (n = 41), Grass-Fed (n = 42), and Wagyu (n = 42)] in order to provide significant variation in beef flavor attributes, tenderness, fat percentages, and animal production. Product specifications for each carcass type were verified by Colorado State University (CSU) personnel using official USDA grades and personal communication with individual suppliers to verify origin.[14] All carcasses selected were A-maturity and had typical beef-type characteristics to avoid dairy-type and *Bos indicus* breed influence. With the exception of Dark Cutter carcasses, all selected carcasses were free from dark cutting lean characteristics. Select, Low Choice, Top Choice, and Prime carcasses presented marbling scores of slight (Sl), small (Sm), modest (Mt), and moderate (Md), and slightly abundant (Slab) and moderately abundant (Mab), respectively. Additional note was taken of these four carcass types meeting the USDA requirements for certification of Angus influence.[15] Grass-Fed carcasses were selected from cattle fed a 100% forage-based diet the entirety of their lives, and carcasses had marbling scores within Sm^{00}–Md^{99} (superscripts represent the numerical marbling score). Wagyu carcasses were selected from crossbred Wagyu cattle (50% Wagyu, 50% Angus) and had marbling scores within $SlAb^{00}$–Mab^{99}. Dark Cutter carcasses were selected from those exhibiting dark cutting characteristics (dark colored lean) with marbling scores from Sm^{00} to Md^{99}. From each half of each carcass, a 5 cm thick strip loin section was collected from a point starting at the 13th rib. Each strip loin section was cut to yield a single 2.54 cm steak (NAMP 1180). The steak from the left side of the carcass was used for REIMS analysis, and the steak from the right side of the carcass was used for shear force analysis. Steaks were individually vacuum-packaged, aged (1°C) for 14 d postmortem, and frozen (–20°C) until analysis.

Shear Force. Slice shear force (SSF) measurements were obtained from every steak using procedures described by Shackelford et al.[16] Within 5 min of recording peak internal temperature, a 1 cm × 5 cm slice was removed from the steak parallel to the muscle fibers from the lateral end and sheared perpendicular to the muscle fibers, using a slice shear force machine (Model GR-152, Tallgrass Solutions, Inc., Manhattan, KS) equipped with a flat, blunt-end blade (crosshead speed: 500 mm/min, load capacity: 100 kg), resulting in a single SSF measurement for each steak. Slice shear force values are reliable objective measurements of meat tenderness and were used to divide the observations into two classifications. According to standards, steaks were classified into two tenderness categories, tender or tough.[17] Steaks with an SSF value less than 20.0 kg were classified as

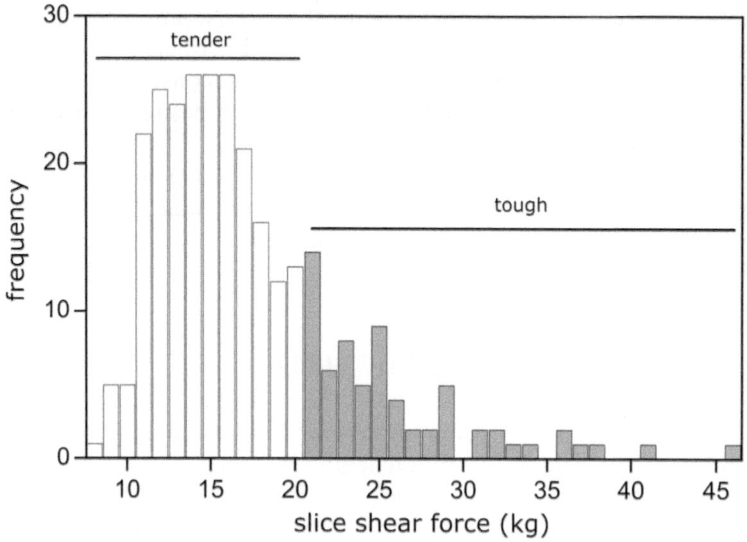

FIGURE 12.1 Histogram of slice shear force values and tenderness classifications as tender (<20.0 kg) and tough (≥20.0 kg).

tender, and steaks with an SSF value equal to or greater than 20.0 kg were classified as tough. Measured SSF values ranged from 8.20 kg to 45.57 kg with an average of 17.33 kg and a standard deviation of 5.97 kg. The distribution of the SSF values coded by tenderness classification is illustrated in Figure 12.1.

Rapid Evaporative Ionization Mass Spectrometry (REIMS). Chemical fingerprints of strip loin sections were acquired using rapid evaporative ionization mass spectrometry (REIMS). Prior to analysis, samples were thawed at 0–4°C for 16–24 h. Samples were analyzed using a Synapt G2 Si Q-ToF, fitted with an REIMS ionization source coupled with a monopolar electrosurgical hand piece ("iKnife", Waters Corporation, Milford, MA) powered by an Erbotom ICC 300 electrosurgical generator (Erbe Elektromedizin GmbH, Tubingen, Germany) used in the "dry cut" mode at a power of 40 W. A continual flow (200 µL/min) of 2 ng/mL leucine-enkephalin was introduced directly to the REIMS source during sampling. The cone voltage was set to 40 V, and the heater bias to 80 V. At least five "burns" were collected for each sample within a 2.54 × 2.54 cm square from the center of the steak (Supplementary Fig. 1), with each burn lasting approximately 1 sec. Spectra were collected in negative mode ionization from 100–1,000 *m/z*. Preprocessing was performed using the LiveID software (Waters Corporation) and includes lock mass correction (leucine-enkephalin), background subtraction using standard Masslynx preprocessing algorithms, and normalization to total ion current. Peak binning was performed at intervals of 0.5 *m/z* resulting in a total of 1,800 bins. The bins from the five burns were summed to create a single value for each sample. Mass bins in the range of 550–600 were excluded from the data matrix to remove the internal standard signal (leucine-enkephalin, *m/z* 554.632) resulting in a final data matrix containing 1,700 variables (*m/z* bins) and 290 observations (samples).

Data Analysis. All data reduction, machine learning, and evaluation of predictive models were performed within the R statistical environment.[18]

Data were grouped together to create the desired classifications for each model set. The classification groups and four model sets are described in Table 12.1 and include

TABLE 12.1 Summary of Classification Groupings and Number of Observations Used for Each of the Four Model Sets

	Model Sets			
Classifications (# of Observations)	**Main**	**Specialized**	**Breed**	**Tenderness**
Dark Cutter (41)	×	×		
Top Choice/ Prime* (81)	×			
Low Choice/ Select* (84)	×			
Wagyu (42)	×	×		
Grass fed (42)	×	×		
Tender (215)				×
Tough (74)				×
Angus (159)			×	
Not Angus (46)			×	

"Main", "Specialized", "Breed", and "Tenderness". Due to similarities in sensory performance (data not shown), Select and Low Choice, as well as Top Choice and Prime samples were combined to create two new classification groups: Low Choice/Select and Top Choice/Prime. The Main model set used the five classification categories: Low Choice/Select, Top Choice/Prime, Dark Cutter, Grass-Fed, and Wagyu. The Specialized model set exclusively used the Wagyu, Grass-Fed, and Dark Cutter classifications. The Breed model set was based on classification of Angus or non-Angus as determined from the carcass meeting the USDA requirements for certification of Angus influence or not.[15] Additionally, only data from Select, Low Choice, Top Choice, and Prime carcasses were used in the Breed model set. Lastly, the Tenderness model set was defined by the slice shear force measurement as described above.

Data Pre-Processing with Dimension Reduction. Dimension reduction was performed using (i) principal component analysis (PCA), (ii) feature selection (FS), or (iii) PCA followed by FS (PCA-FS). PCA dimension reduction was performed using the *PCA* function in the package *FactoMineR* with unit variance scaling.[19] FS is a supervised method of data reduction where the predictors chosen are specific to the separation of the given classes. FS was performed separately for each model set in the study (i.e. Main, Specialized, Breed, and Tenderness) using the *caret* R package.[20, 21] Data were preprocessed by removing highly correlated *m/z* bins (Pearson's $|r| > 0.90$; *findCorrelation* function) followed by the *rfe* function and finally assessed with five-fold cross validation. During the feature selection process, all 1,700 *m/z* bins were considered to account for all possible selected variables. PCA-FS consisted of performing a similar feature selection process on the principal components, rather than the 1,700 mass bins. PCA-FS was performed using ten-fold cross validation because it was significantly more computationally efficient to conduct FS on the principal components than on the entire set of 1,700 *m/z* bins. For all model assessment performed in this study, X-fold cross validation refers to removal of (100/X)% of the data as a validation set where the remaining (100 – (100/X))% is used as the training set. This procedure was repeated X times, and the average of the prediction accuracy is recorded.

Machine-Learning Algorithms to Predict Beef Quality. A total of eight machine-learning algorithms were compared for predictive accuracy of each model set. These include: (i) support vector machine with a linear kernel (SVM-L), radial kernel (SVM-R), and polynomial kernel (SVM-P), (ii) random forest (RF), (iii) K-nearest neighbor (Knn),

(iv) linear discriminant analysis (LDA), (v) penalized discriminant analysis (PDA), (vi) extreme gradient boosting (XGBoost), (vii) logistic boosting (LogitBoost), and (viii) partial least squares discriminant analysis (PLSDA). An initial screening of all the machine-learning algorithms except PLSDA was performed using the *train* function in the *caret* package. PLSDA is not supported in the *train* function, and thus PLSDA models were constructed using the *plsDA* function[22] built into the *DiscriMiner* package. Visualization plots of the PLSDA model based on the PCA-FS reduced data were generated using the *ggpubr* package. Ellipses were drawn using a 90% probability region. Ten-fold cross validation was used to evaluate the prediction accuracy (correct predictions/total predictions × 100) of all eight machine-learning algorithms. For PLSDA, a manual ten-fold cross-fold validation was conducted using the *predict* function.

For each of the four model sets (Main, Specialized, Breed, Tenderness), the eight machine-learning algorithms were applied on data following the four preprocessing options (no reduction, PCA, FS, or PCA-FS reduction). Prediction accuracies of each model were determined by ten-fold cross validation.

The top three performing models (in terms of prediction accuracy based on ten-fold cross validation) for each model set were further optimized via parameter tuning using the specific functions in R rather than the *train* function (Table 12.2). Following optimization, the top three models were revalidated using 100-fold cross validation for each model set (a higher fold cross validation was used to more precisely estimate the precision accuracy).

12.2 RESULTS AND DISCUSSION

REIMS analysis of the beef samples in this study resulted in a data matrix that included 1,700 *m/z* bins per sample, for a total of 290 samples in the full experiment. This type of unbalanced data, typical in omics experiments (i.e. many more predictors than observations), can lead to issues when applying machine-learning algorithms such as slow computational time and the possibility of model overfitting.[13] To address this challenge, three types of dimension reduction methods were compared in this study including principal component analysis (PCA), feature selection (FS), and PCA followed by FS (PCA-FS). In PCA, the data transform the predictor variables into a smaller number of variables (termed principal components, PCs), based on predictor co-variation (direction and magnitude) in the data, and is an unsupervised method.[20] When employed for data reduction, all of the PCs are retained and the output is a new data set that represents 100% of the variation in the data using substantially fewer predictors. In FS, a supervised, recursive

TABLE 12.2 R Packages and Functions Used for the Development of Final Predictive Models

Machine-Learning Algorithm	R Function	Package
LDA	lda()	MASS
LogitBoost	LogitBoost()	*caTools*
PDA	fda ()	mda
SVM	svm()	*e1071*
XGBoost	xgboost()	*xgboost*
PLSDA	plsDA()	DiscriMiner

TABLE 12.3 Number (Percent) of Predictors for Each Dimension Reduction Technique for Each Model Set

Model Set	Original	PCA	FS	PCA-FS
Main	1,700 (100%)	289 (17%)	229 (13.47%)	24 (1.41%)
Specialized	1,700 (100%)	124 (7.29%)	60 (3.53%)	8 (0.47%)
Breed	1,700 (100%)	203 (11.94%)	602 (35.41%)	38 (2.24%)
Tenderness	1,700 (100%)	289 (17%)	67 (3.94%)	16 (0.94%)

feature elimination method is used to remove predictor variables from the data matrix.[20] FS applies a backwards selection of predictors based on a ranking of predictor importance. The less important predictors are sequentially eliminated, with the goal of finding the smallest subset of predictors that can generate an accurate predictive model. The PCA-FS method is based on reducing the data into components, then performing FS on components to reduce redundancy in the data matrix.

Here, the PCA, FS, and PCA-FS methods were applied to the data matrix of 1,700 *m/z* bins for each of the four model sets. The PCA, FS, and PCA-FS methods reduced the data matrix to 226 (13.3%), 240 (14.0%), and 22 (1.3%) mean predictor variables, respectively (Table 12.3). The number of predictors in the processed data was most variable for the FS method (coefficient of variation 92%; among the four model sets), indicating that PCA is likely to produce a more consistently sized data matrix for predictive modeling.

Machine-Learning Algorithms Varied in Their Prediction Accuracy. In this study, eight machine-learning algorithms were compared for the prediction of specific quality attributes in beef based on molecular profiles generated by REIMS.

Partial least squares discriminant analysis (PLSDA). A model that transforms data into partial least squares components that can be then used to classify an observation. PLSDA is a common chemometrics method used in mass spectrometry "omics" experiments to predict outcomes based on chemical signals;[13, 23] however this algorithm can be easily misused, misinterpreted, and is prone to overfitting.[23] The PLSDA algorithm has many advantages including performance power when working with multivariate data along with methods of dealing with collinear variables.[24] PLSDA is similar to a supervised version of principal component analysis. However, in this study, the PLSDA model was used as a classification algorithm and for data visualization rather than as a dimension reduction technique. The data were visualized using the first two PLS components of the PCA-FS dimensionally reduced data (Figure 12.2). The Specialized model set showed clear separation between classes, with slight overlap between the Dark Cutter and Grass-Fed classes. The most overlap was observed in the Main model set where only the Wagyu class appears to separate clearly from the others. For the Breed and Tenderness model sets, there is clear separation of the classes with some overlap of the moderate values.

Support vector machine (SVM). A discriminative classifier that separates clusters of observations with the use of a hyperplane. In particular, the SVM method determines the optimal hyperplane to differentiate between the classes within the data. Here, this study evaluated how the "kernel" parameter of SVM (linear, radial, or polynomial) can result in varying prediction accuracy.

Random forest (RF). A type of decision tree. Although individual classification trees tend to lack in performance compared to other machine-learning algorithms, aggregating many decision trees together with methods such as bagging, random forests, and boosting

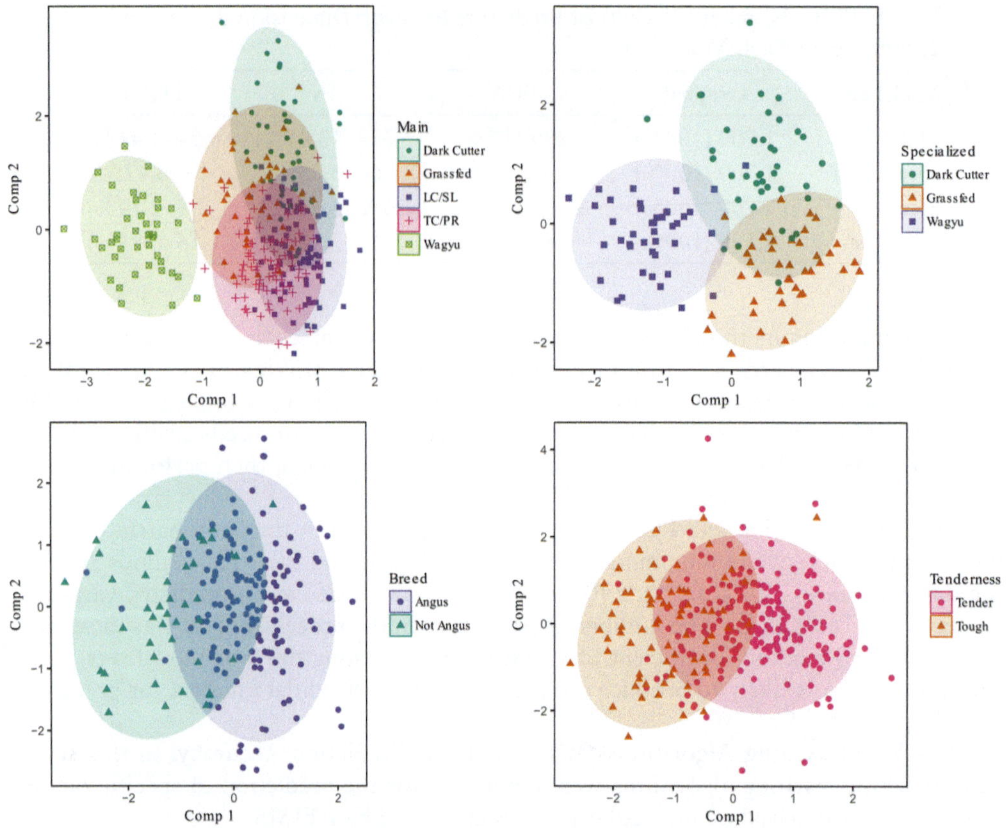

FIGURE 12.2 Visualization of the PLSDA model for each of the model sets. Plots represent the first two PLS components of the PCA-FS reduced data.

can greatly increase the predictive accuracy of the model.[25] The RF algorithms can increase model performance compared to other classification tree methods by decorrelating the trees.[25] Similar to bagging, random forest methods construct *n* number of decision trees on bootstrapped training samples. However, the unique component of the random forest model is that for each decision split within a given tree, the model is only able to use *m* predictors. The parameter *m* can be set as any value less than or equal to the number of predictors (note that if *m* equals the number of original predictors it would be performing a bagging classification method). Typically, *m* is chosen to be approximately equal to the square root of the number of predictors. By only using *m* predictors at each split, the trees are decorrelated due to the same predictors not being selected for each of the trees.

K-nearest neighbor (Knn). A nonparametric approach with no underlying assumption about the distribution of the data. This algorithm classifies observations based on the similarities in features between individuals. The model determines feature similarity by calculating the Euclidian distance between the features of different observations and assigns a distance value to each observation and its neighboring observations.[25] Deciding the best K value for a given data set is an optimization problem. A loop can be used to input various K values into the algorithm to find the value of K that minimizes the error rate of class prediction.

Linear discriminant analysis (LDA). A parametric approach that assumes the predictors $X_1, \ldots, X_k$ are drawn from a multivariate Gaussian distribution. LDA is a mathematically simple and robust method of classification. LDA uses linear decision boundaries for the classification of observations, and this method calculates a linear combination of predictors to separate the model's classes.[25]

Penalized discriminant analysis (PDA). An expansion of the linear discriminant analysis model. The PDA algorithm uses nonlinear spline basis functions and includes a penalty term that adds smoothness to the coefficients of the model to reduce the problem of multicollinearity in the predictors.[26] Therefore, this penalized algorithm typically performs well when there are many highly correlated variables. When there are a large number of correlated variables within a data set, many times, the covariance data matrix is non-invertible.[25] Including a penalization parameter in the model reduces the likelihood of singular (non-invertible) covariance matrices and results in improved classification accuracy.

XGBoost. A supervised learning algorithm designed for fast computational time, especially on very large data sets. XGBoost is a form of gradient-boosted decision trees that is faster than comparable implementations of gradient boosting. Gradient boosting can generate new models based on the prediction of the residual errors of prior models.[27] The term "gradient boosting" refers to the utilization of a gradient descent to minimize the loss when adding additional models (Brownlee, 2016).

LogitBoost. A boosting logistic classification algorithm that performs as an additive logistic regression model. The LogitBoost model is similar to a generalized additive model, but rather than minimizing the exponential loss, the algorithm minimizes the logistic loss of the function. Additionally, the LogitBoost algorithm in R is trained using one-node decision trees as weak learners.[28]

The performance of each machine-learning algorithm and data reduction combination was assessed in the initial screening step (Supplementary Figs 1–6). Performance was evaluated in terms of prediction accuracy using a ten-fold cross validation. The best performing machine-learning algorithm and data reduction combinations for each model set are summarized in Figure 12.3 and Supplementary Table 1. For the two binary model sets, Breed and Tenderness, the prediction accuracies among the highest performing machine-learning algorithm data reduction approach combinations to the lowest span only 4.5% and 9.4%, respectively (Breed range: 0.78–0.825; Tenderness range: 0.814–0.908). This result supports that all of the approaches generated a consistently accurate model. More variation was observed in the prediction accuracies for the complex Main and Specialized model sets with predictions accuracies spanning 22.7% and 24%, respectively (Main range: 0.536–0.763; Specialized range: 0.728–0.968).

Parameter tuning and optimization were performed for the top three machine-learning algorithms for each model set. The overall highest performing machine-learning algorithm and data reduction approach combination for each model set was selected based on prediction accuracy (100-fold cross validation) using the optimized algorithms. Prediction accuracies for the highest performing models are reported in Table 12.4. Interestingly, the most accurate machine-learning algorithms were different for each model set, specifically LDA, XGBoost, and SVM (both linear and radial), although in many cases prediction accuracies of the top-ranked algorithms differed by less than 1% (Figure 12.3). Additionally, dimension reduction using either FS or PCA-FS was optimal for all of the top-performing algorithms.

Relevance of Machine Learning of REIMS Data for Beef Quality Predictions. The results obtained in this study demonstrate that integrating machine learning with REIMS data can predict beef quality attributes with considerable accuracy, including quality grade, production background, breed type, and muscle tenderness. Dimension reduction improved the predictive accuracy in all cases, supporting that this is a critical

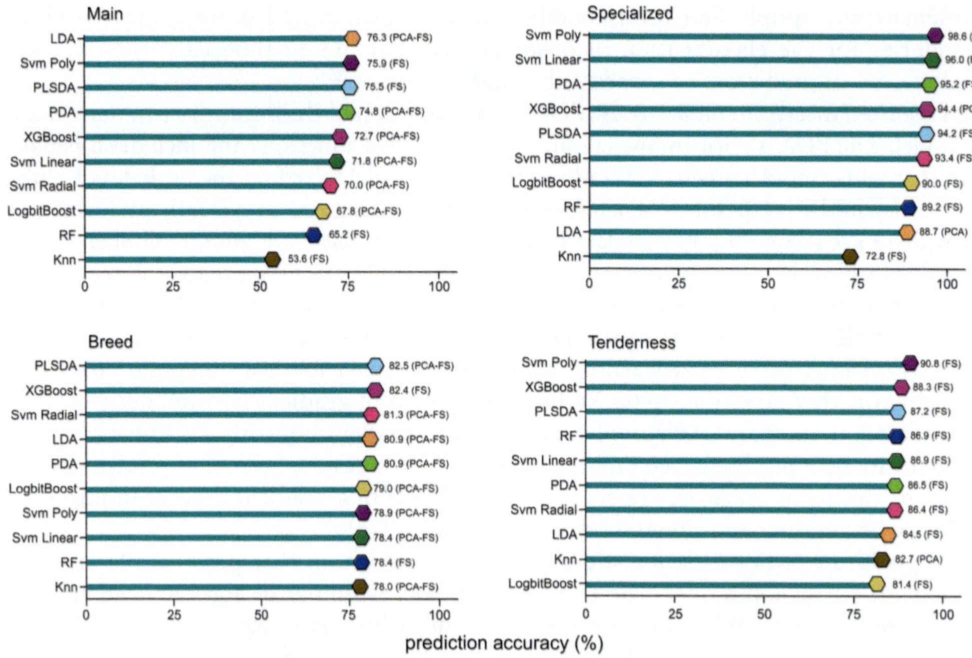

FIGURE 12.3 Prediction accuracies (based on ten-fold cross validation) for the top-performing machine-learning algorithm and data reduction approach combinations for each model set.

step in data processing and analysis. Further, machine-learning algorithms varied in their performance depending on the model set, indicating that the patterns in the chemical data (i.e. REIMS spectra) are highly complex and variable for the different facets of beef quality attributes. Thus, finding a "one-size-fits-all" approach to generating predictive models for beef quality attributes is unlikely, and instead, evaluation of multiple algorithms should be standard practice in model development with highly complex chemical data.

Our results support the potential for REIMS analysis to be further developed to complement the beef quality classification systems. Tenderness is a critical attribute for consumer satisfaction, which can exist independently of USDA quality grade, breed type, or production background.[29–31] Slice shear force can be used to verify guaranteed tender programs, but this method is laborious, costly, and destructive, and the industry has not

TABLE 12.4 Summary of Final Prediction Accuracies Based on 100-Fold Cross Validation for the Top Machine-Learning Algorithm and Data Reduction Approach Combination for Each Model Set after Parameter Tuning

Model Set	Dimension Reduction Approach	Number of Predictors	Machine-Learning Algorithm	Final Accuracy Rate
Main	PCA-FS	24 PCs	LDA	**81.5%**
Specialized	FS	60 mass-bins	SVM - Linear	**99%**
Breed	PCA-FS	38 PCs	SVM - Radial	**85%**
Tenderness	FS	67 mass-bins	XGBoost	**90.5%**

widely adopted its use for individual carcass classification for product labeling.[32] Several instrument methods to classify beef tenderness have been evaluated that are less destructive than SSF and could be implemented at line speeds, but have yet to be adopted by the industry for routine use[33–38] In this study, REIMS output, coupled with machine learning, correctly classified tough and tender carcasses with more than 90% accuracy (Table 12.1), indicating the potential value of this approach for industry use.

Similarly, beef from Angus breed type cattle can receive significant premiums, as it is a requirement for several of the most desirable and highest quality branded beef programs. However, Angus influence is most commonly determined by visually assessing the predominance of black coloring of the live animal's hide, rather than a true genetic test or physical documentation of lineage. Several machine-learning algorithms evaluated in this study predicted Angus breed type with greater than 80% accuracy. It is important to note that this result does not represent a prediction of genuine Angus influence, but rather a prediction of a carcass originating from an animal with a predominantly black-colored hide. However, the results support the potential for prediction of true Angus genetic influence in future work, where carcasses with known genetic background could provide a more objective alternative to identifying Angus influence. Successful prediction of Grass-Fed, Wagyu, and Dark Cutter carcasses with considerable accuracy in the current study suggests additional potential to utilize REIMS in determining and/or verifying various quality-related beef carcass characteristics.

12.3 CONCLUSIONS

The current study demonstrates that chemical profiles generated by REIMS and interpreted with machine learning algorithms can generate predictive models for beef quality attributes such as carcass type, production back- ground, breed type, and muscle tenderness. The optimal machine-learning algorithm, as assessed by predictive accuracy, was different depending on the classification problem, suggesting that a "one-size-fits-all" approach to developing predictive models from REIMS data is not optimal. Furthermore, in all cases, data reduction prior to modeling improved the overall model accuracy. Taken as a whole, the results presented here lay the groundwork for future evaluation of REIMS in an on-line production setting to complement current meat classification methodologies and enable objective sorting and verification of meat products by attributes with high economic value.

DATA AVAILABILITY

Raw mass spectrometry data files have been uploaded to the Metabolights data repository. Preprocessed mass spectrometry data and additional metadata for each sample are tabulated in Supplementary Data Table.

REFERENCES

1. Black, C., Chevallier, O. P. & Elliott, C. T. The current and potential applications of Ambient Mass Spectrometry in detecting food fraud. *TrAC Trends in Analytical Chemistry* **82**, 268–278 (2016).

2. Takats, Z., Wiseman, J. M., Gologan, B. & Cooks, R. G. Mass spectrometry sampling under ambient conditions with desorption electrospray ionization. *Science* **306**, 471–473 (2004).
3. Weston, D. J. Ambient ionization mass spectrometry: current understanding of mechanistic theory: analytical performance and application areas. *The Analyst* **135**, 661–668 (2010).
4. Schafer, K. C. *et al. In vivo*, *in situ* tissue analysis using rapid evaporative ionization mass spectrometry. *Angewandte Chemie (International ed. in English)* **48**, 8240–8242 (2009).
5. Kinross, J. M. *et al.* iKnife: rapid evaporative ionization mass spectrometry (REIMS) enables real-time chemical analysis of the mucosal lipidome for diagnostic and prognostic use in colorectal cancer. *Cancer Research* **76** (2016).
6. St John, E. R. *et al.* Rapid evaporative ionisation mass spectrometry towards real time intraoperative oncological margin status determination in breast conserving surgery. *Cancer Research* **76** (2016).
7. Phelps, D. *et al.* Diagnosis of borderline ovarian tumours by rapid evaporative ionisation mass spectrometry (REIMS) using the surgical intelligent knife (iKnife). *Bjog-International Journal of Obstetrics and Gynaecology* **123**, E4–E4 (2016).
8. Balog, J. *et al.* Identification of the species of origin for meat products by rapid evaporative ionization mass spectrometry. *Journal of Agricultural and Food Chemistry* **64**, 4793–4800 (2016).
9. Black, C. *et al.* A real time metabolomic profiling approach to detecting fish fraud using rapid evaporative ionisation mass spectrometry. *Metabolomics* **13**, 153 (2017).
10. Guitton, Y. *et al.* Rapid evaporative ionisation mass spectrometry and chemometrics for high-throughput screening of growth promoters in meat producing animals. *Food Additives and Contaminants* **35**, Epub 2018 Jan 17 (2017).
11. Verplanken, K. *et al.* Rapid evaporative ionization mass spectrometry for high-throughput screening in food analysis: the case of boar taint. *Talanta* **169**, 30–36, Epub 2017 Mar 21 (2017).
12. Smith, T. C. & Frank, E. In *Statistical Genomics: Methods and Protocols* (eds Ewy Mathé & Sean Davis), pp. 353–378 (Springer, New York, 2016).
13. Gromski, P. S. *et al.* A tutorial review: metabolomics and partial least squares-discriminant analysis–a marriage of convenience or a shotgun wedding. *Analytica Chimica Acta* **879**, Epub 2015 Feb 11 (2015).
14. United States Department of Agricutlure, A. M. S. *United States Standards for Grades of Carcass Beef.* (2017).
15. United States Department of Agricutlure, A. M. S. *Specification for Characteristics of Cattle Eligible for Approved Beef Programs Claiming Angus Influence.* (2016).
16. Shackelford, S. D., Wheeler, T. L. & Koohmaraie, M. Evaluation of slice shear force as an objective method of assessing beef longissimus tenderness. *Journal of Animal Science* 77, 2693–2699 (1999).
17. ASTM. *Standard Specification for Tenderness Marketing Claims Associated with Meat Cuts Derived from Beef* (West Conshohocken, PA, 2018).
18. Team, R. C. *R: A Language and Environment for Statistical Computing*, http://www.R-project.org/ (2013).
19. Francois, H. *et al. Exploratory Multivariate Analysis by Example Using R* (Chapman and Hall/CRC, 2018).
20. Kuhn, M. & Johnson, K. *Applied Predictive Modeling*, 1 edn (Springer-Verlag, 2013).

21. Kuhn, M. Building Predictive Models in R Using the caret Package. *J Stat Softw* (2008).
22. Pérez-Enciso, M. & Tenenhaus, M. Prediction of clinical outcome with microarray data: a partial least squares discriminant analysis (PLS-DA) approach. *Human Genetics* **112**, 581–592 (2003).
23. Gromski, P. S. *et al.* A comparative investigation of modern feature selection and classification approaches for the analysis of mass spectrometry data. *Analytica Chimica Acta* **829**, 1–8 (2014).
24. Perez, D. R. & Narasimhan, G. So you think you can PLS-DA? *bioRxiv* (2018).
25. Adler, A. E. *et al.* Comparing classification methods for diffuse reflectance spectra to improve tissue specific laser surgery. *BMC Medical Research Methodology* **14**, 91 (2014).
26. Lenz, R. & Decker, H. J. *Advances in Data Analysis* (Springer, 2018).
27. Chen, T. & Guestrin, C. XGBoost: a scalable tree boosting system. In *Proceedings of the 22nd ACM SIGKDD International Conference on Knowledge Discovery and Data Mining*, pp. 785–794 (ACM, San Francisco, CA, 2016).
28. Tibshirani, J. F., Trevor, H. & Robert, T. Additive logistic regression: a statistical view of boosting (with discussion and a rejoinder by the authors). *The Annals of Statistics* **28**, 337–407 (2000).
29. Boleman, S. J. *et al.* Consumer evaluation of beef of known categories of tenderness. *Journal of Animal Science* **75**, 1521–1524 (1997).
30. Miller, M. F. *et al.* Consumer thresholds for establishing the value of beef tenderness. *Journal of Animal Science* **79**, 3062–3068 (2002).
31. Platter, W. J. *et al.* Effects of marbling and shear force on consumers' willingness to pay for beef strip loin steaks. *Journal of Animal Science* **83**, 890–899 (2005).
32. Wheeler, T. L. *et al.* The efficacy of three objective systems for identifying beef cuts that can be guaranteed tender. *Journal of Animal Science* **80**, 3315–3327 (2003).
33. Park, B. *et al.* Near-infrared reflectance analysis for predicting beef longissimus tenderness. *Journal of Animal Science* **76**, 2115–2120 (1998).
34. Belk, K. E. *et al.* Evaluation of the Tendertec beef grading instrument to predict the tenderness of steaks from beef carcasses. *Journal of Animal Science* **79**, 688–697 (2001).
35. Vote, D. J. *et al.* Online prediction of beef tenderness using a computer vision system equipped with a BeefCam module. *Journal of Animal Science* **81**, 457–465 (2003).
36. Konda Naganathan, G. *et al.* Partial least squares analysis of near-infrared hyperspectral images for beef tenderness prediction. *Sensing and Instrumentation for Food Quality and Safety* **2**, 178–188 (2008).
37. Konda Naganathan, G. *et al.* Three dimensional chemometric analyses of hyperspectral images for beef tenderness forecasting. *Journal of Food Engineering* **169**, 309–320 (2016).
38. Nubiato, K. E. Z. *et al.* A bench-top hyperspectral imaging system to classify beef from Nellore cattle based on tenderness. *Infrared Physics and Technology* **89**, 247–254 (2018).

ACKNOWLEDGMENTS

The authors would like to thank the Beef Checkoff for funding. We would also like to thank Dr. Julia Balog for her input and assistance with REIMS data preprocessing.

AUTHOR CONTRIBUTIONS

D.A.G., D.A.K., S.D.S., D.R.W., K.E.B., and T.L.W. were responsible for sampling design and sample collection; D.A.G. performed S.S.F. measurements; D.A.G., C.D.B., and J.E.P. participated in data acquisition; D.A.G., A.R.C., S-Y.K., and J.L.S. performed data analysis and statistics; J.E.P., D.A.G., A.R.S., and A.L.H. wrote manuscript text; all authors reviewed and edited the manuscript.

ADDITIONAL INFORMATION

Supplementary information accompanies this paper at https://doi.org/10.1038/s41598-019-40927-6.

Competing interests: The authors declare no competing interests.

Index